PRAISE ... JESSICA STERN

"[A] stunningly brave book." —*Elle*

"A gutsy investigation into the ways in which trauma is perpetuated." —*Vogue*

"[Stern's] commitment to introspection makes for a book that is memorably searing." —*New York Times Book Review*

"[*Denial* is a] powerfully constructed memoir . . . [an] incandescently honest book." —*Washington Post*

"[Stern's rapist], whose chaotic life and whose own probable victimization Stern reconstructs, caused her lifelong anguish. She doesn't simply tell us so but shows us in shattered, artfully repetitive narration." —*Boston Globe*

"An intense and honest examination. . . . *Denial* is a touching portrait of how the after affects of trauma can influence the personal and professional life of its victim." —*Philadelphia Tribune*

"[Stern's] eloquent, vital book is mesmerizing. Long before I finished, I had made a mental list of several friends on whom I will press it, knowing that it will trigger the kind of discussion one has with those one trusts the most. . . . With *Denial*, Jessica Stern has written a brilliant, indispensable book."

—*Plain Dealer* (Cleveland)

"An extraordinary memoir conveys Stern's process of denial, dissociation, and healing in her dawning realization of intolerable truths." —*Providence Journal*

"In this skillfully wrought, powerful study, a terrorism expert, national security adviser, and lecturer at Harvard returns to a definitive episode of terror in her own early life and traces its grim, damaging ramifications. . . . Stern's work is a strong, clear-eyed, elucidating study of the profound reverberations of trauma."

—*Publishers Weekly* (starred review)

"Wonderfully compassionate, absorbing reading for anyone."

—*Booklist* (starred review)

"Most moving is the author's contemplation of denial itself, and its effect of re-victimizing the victim. . . . She successfully unearths difficult emotional terrain without sinking into utter subjectivity and maintains an orderly progression without becoming clinical. A disturbing, captivating memoir." —*Kirkus Reviews*

"A masterpiece. A remarkable human journey from confusion and doubt to clarity and perspective. Stern gives an incisive account of the shape of the imprint of trauma on body and soul, and shows us how honest confrontation with what we already know, but try to forget, is essential in order to be liberated from the past."

—Dr. Bessel van der Kolk, M.D., Professor of Psychiatry,
Boston University Medical School, and founder
and medical director of the Trauma Center

"An unflinchingly courageous self-examination of the impact of trauma on an individual's unfolding life. Through her riveting and brilliantly told story, Jessica Stern lays out the impact of trauma on an individual's ways of experiencing the world, how trauma is transmitted across generations, and how severe trauma can be adaptively transformed. The book will be illuminating for victims and survivors of trauma, those who work or live with them, family members with generational histories of trauma, and for those who care about how our histories shape our lives."

—Edward R. Shapiro, M.D., medical director and CEO,
Austen Riggs Center, and Clinical Professor of Psychiatry,
Yale University School of Medicine

"Jessica Stern has written a remarkable book, unlike any that I've read. This deeply personal and often painful reflection documents the costs of personal, familial, and community silence as well as the liberating effects of truthful testimony."

—Howard Gardner, Hobbs Professor of Cognition and Education at the Harvard Graduate School of Education and author of *Multiple Intelligences*

"Jessica Stern's harrowing memoir of a girl whom trauma has taught to distrust herself and who learns to live with the idea of her helplessness—a girl who once turned away from what she could not understand or accept—examines the violence at the heart of things, with an appeal to compassion and forgiveness, rather than a condemnation of the destructive impulses that haunt each of us."

—Susanna Moore, author of *In the Cut*

"A memorable, powerful, and deeply courageous book, *Denial* is also a riveting read. Stern brings a unique combination of insight, empathy, and acumen to bear in helping us to understand the perpetrators of violence. With devastating honesty she explores the impact of trauma on victims and those close to them, and the costs of denial for both."

—Louise Richardson, author of *What Terrorists Want*

"Jessica Stern has always had the gift of disappearing into the lives and minds of terrorists. In this book, she faces a greater challenge of her own. The layers of abuse she has encountered, her tender matter-of-factness, her refusal of self-pity in favor of insatiable curiosity—these are some of her gifts born of trauma. This book will allow people into parts of themselves they did not have access to, or even know they had. Parts full of rage, of terror, of pride in their own detachment. It will allow their hearts to begin to break. For anyone who has lived at proximity to violence, it is one of the most necessary accounts of our time."

—Eliza Griswold, journalist and author of *Wideawake Field*

"Extraordinary. I am speechless. This book is truly important. It will change—at least, help heal—lives."

—Susan Willet Bird, founder and CEO of Wf360, the creators of *Brandversation*

—[illegible], poet, novelist and author of *[illegible]*

denial

ALSO BY JESSICA STERN

Terror in the Name of God

The Ultimate Terrorists

denial

A MEMOIR OF TERROR

jessica stern

An Imprint of HarperCollinsPublishers

All letters courtesy of the authors.

"I Am Lucy" courtesy of Amy Vorenberg.

A hardcover edition of this book was first published in 2010 by Ecco, an imprint of HarperCollins Publishers.

FIRST ECCO PAPERBACK EDITION PUBLISHED 2011.

Library of Congress Cataloging-in-Publication Data has been applied for.

ISBN 978-0-06-162666-1

HB 10.28.2022

author's note

Knowing how traumatic reliving these times was for me, I did not want to expose the people who spoke to me about their own experiences. So, for many of the subjects I interviewed—victims, family, friends, and law enforcement officials—I have changed names and modified other identifying characteristics, such as their towns of residence, in order to preserve their anonymity. In chapter five, the words I attribute to Mary (a pseudonym) were actually combined with those of another person who knew the rapist. I have also chosen to change the names and other characteristics of the many priests who were named as possible abusers. Even though many of them were cited in the criminal and civil complaints that overwhelmed the Worcester and Boston dioceses in past years, confirming their connection to these events was beyond the scope of my project. The goal in all cases was to protect people's privacy without damaging the integrity of the story.

This book is dedicated to my father,
who taught me the value of persistence,
and that courage comes in many guises

contents

acknowledgments

I would like to thank Jason Epstein for proposing that I write this book. It took me some time to understand that the subject would be of interest to others, but Jason was right, as usual. I would not have had the courage to start, or finish, without the early and continuing support of Jerry Fromm, Howard Gardner, Hillary Chute, James Gilligan, Jack Goldsmith, Robert J. Lifton, Catherine MacKinnon, Julia Moore, Louise Richardson, Ed Shapiro, and Ben Wittes. My sister, Sara Stern Fishman, revisited the terrors described in this book with me. I am sorry that in writing this book, I inadvertently exposed her to the memory of these horrors, and I am grateful, as always, to have her hand to hold.

I am deeply grateful to my agent, Martha Kaplan, who intuited there was a book in this material, long before there was empirical evidence, and for her constant encouragement. Martha holds you up when you most need holding. I thank my editors at Ecco, Dan Halpern and Ginny Smith. Ginny's probing questions, delivered with grace, make clear that she will go far as an editor. Ben Wittes and Dan Halpern both insisted that I cease hiding behind policy prescriptions and simply tell the story.

The book was supported by grants from the Guggenheim Foundation and the Women and Public Policy Program at the Kennedy School of Government. I benefitted also from residencies at the Erik Erikson Institute, MacDowell Colony, and Yaddo. I am deeply grateful for the support of these organizations and the individuals involved. The Erik Erikson Institute offered me an opportunity to present chapters in draft. I learned a great deal from the comments of participants. I am grateful, too, for the time I was able to spend writing in silence on the fifth floor of the Boston Athenaeum.

Barbara Pizer, my very dear friend, was the kind of reader writers dream about. She knew how to read, not just my words, but also my intentions. She persuaded me that I had it in me to go further, both in life and on the page.

A number of government officials helped by digging up documents in archives or sharing their memories. I would like especially to thank Paul Macone, who is already a hero in Concord, Massachusetts, but deserves to be much more widely known. His integrity, diligence, and sensitivity make him a model public official.

I have been assisted by extraordinary research assistants. Jack McGuire is a world-class sleuth. He didn't just find documents for me, but people whom no one else could find. I would also like to thank Desmond O'Reilly. I am in awe of Desmond's determination, strength, and faith. I would like also to thank Brooke McConnell.

My dear husband, Chet Atkins, learned to recognize when terror was influencing my actions, long before I understood to recognize this feeling myself. He drove me to every interview, and listened to every word. This book, in many ways, is a testament to my desire, inspired by him, to learn to love.

This book is written for victims—raped or traumatized or terrorized—many of whom outsiders may not recognize under their soldiers' uniforms or courageous façades, but who may recognize themselves in the pages that follow. I hope this book provides comfort and hope.

preface

For the last twenty years, I have been studying the causes of evil and violence. Until now I never questioned why I was interested in this work, or why I was able to do it. This book answers a question that is always addressed to me when I speak about my work on terrorism: How could a "girl" like you visit terrorist training camps in Pakistan? Weren't you afraid? The answer is that I wasn't aware that I was afraid—and this book explains why. After a series of traumas, one can lose the capacity to feel fear appropriately.

My narrative takes at its starting point the hour that a rapist spent with a gun trained on my sister and me when she was fourteen and I was a year older. Both my sister and I went on to lead relatively happy and productive lives after the trauma. My sister is a successful marketing communications executive, an opera singer, and an actress. She is married and has two children. She feels great joy in her family and in her music, and no one would describe her as a victim. I similarly take enormous pleasure from my family and my work.

And yet, from childhood on, I noticed puzzling changes that seemed to grow worse over time. With each passing year, I seemed to feel less and less—less pain, but also less joy. As a child, I wanted to be a writer, but continuing bad grades in classes that required writing persuaded me to give up. I found myself more comfortable studying unemotional subjects. Instead, I majored in chemistry, in part because it came relatively easily to me, and in part because I found it comforting that the answers were either right or wrong, unlike in real life, where emotional valences count. I was planning to become a chemist, but then I got seduced by curiosity about violence. I was both repulsed and fascinated. I skipped the war parts in *War and Peace*, but wrote a doctoral dissertation on chemical weapons, which focused mainly on the mechanics of violence, with little attention to the human toll.

Ultimately, I became an expert on terrorism. I wrote my first article on terrorism in 1983. At the time, this was an eccentric choice, and not a very wise career move. Very few people took the threat of terrorism seriously. Still, I intuited that terrorism would become increasingly important, and I continued working on it. I started out doing technical work, related to terrorist weaponry. But eventually I gave in to an intense curiosity about the terrorists themselves. In that work I made use of a personality quirk, rather than my academic training. I am fascinated by the secret motivations of violent men, and I'm good at ferreting them out.

My unusual reaction to fear turned out to be an asset in this work: I discovered that I could do things that other people find frightening, such as traveling to Beirut or Lahore to meet with terrorists. I found that I was able to silence judgment as I listened, to stop myself from feeling fear or horror. I saw that if I allowed myself to feel only curiosity and empathy—not to be confused with sympathy—the terrorists would want to talk to me.

Nonetheless, fear found me. Situations that other people don't find frightening, such as the sound of fireworks, alarm me inordinately. I do not like to be in crowds, especially at night, when there might be a frisson of sexuality in the air. I do not like to be in shopping malls, especially under fluorescent lights. When I am nervous for any reason, I find the clicking of a turn signal or the ticking of a clock almost unbearably agitating. But in truly frightening situations, I retain my composure. For example, when I found myself facedown on the floor in the middle of an armed robbery, a peculiar sense of calm enveloped me. I paid little attention to these surprising states of calm. I could not will myself to respond to danger in any other way, and afterward I could hardly remember the incidents that elicited them. I assumed that my unusual response to danger was a personality quirk, unique to me. I had no idea that these peculiar reactions—calm in the face of danger but fear in response to innocuous sounds or scents—were well-studied aftereffects of trauma.

I consulted a therapist—not about my lack of feeling, but because I wanted to feel *less*. I did not consider it strange that I told the therapist that I wanted two things from treatment: to feel less emotion and to be more efficient at work. It seemed to me that feelings—any feelings—got in the way of life. She told me that some of the qualities I assumed I had been born with—including heightened sensitivity to sudden movements, scents, sounds, and light—were actually markers of trauma. She suggested that I might have post-traumatic stress disorder (PTSD).

I did not believe this diagnosis. I knew about PTSD from having worked with soldiers, and I could not imagine that my own life experience would result in similar symptoms. I had long ago brushed the memory of my rape aside. I considered myself to have overcome it, to have "moved on." I wanted to contribute to society rather than remain stuck in the past, cringing in terror. I did not want to allow fear to influence my decisions in any way. And yet, ironically, terror became my central preoccupation. I felt compelled to understand the deeper motivations of those who hurt others. Instead of feeling terror, I studied it.

After the completion of my second book on terrorism, I found myself wanting to understand what had happened to me during and after my rape. I requested a complete copy of the police report. When the police read the file in 2006 in order to redact it, they realized that a child rapist might still be at large. Pedophiles grow old like the rest of us, but they often continue committing the same crimes unless they are physically stopped.

The police required my help. But I also required theirs. Just as I needed to understand the motivations of the terrorists I interviewed, I found myself needing to understand my rapist, as a way to tame a terror that I was beginning to feel for the first time.

I soon realized that I had forgotten many of the details of the rape, even though I was not a small child when the crime occurred. I felt compelled to answer a question I have spent my professional life asking in regard to terrorists: What happened to the boy who became my rapist? Was there anything in his life story that might explain, at least in part, why he would want or need to hurt my sister and me? What happened to him afterward? I also wanted to know: Why did I go into a trancelike state during the rape, and why have I continued to go into that state whenever I feel truly threatened? And there were larger questions that pulled me in: Why does the threat of violent death alter some of us, even if subtly, forever? Why does it make us unusually numb or calm when we ought to feel terrified? Why do scents or sounds trigger in some of us a feeling of terror or unbearable dread, even in situations where we know, at least intellectually, that we are perfectly safe?

The story turned out to be much bigger than the rape of two girls. It seemed as if the entire community was in denial. The police had not properly investigated the crime. They gave up quickly. They did not believe my sister and me when we insisted that we did not know the perpetrator. And rape at gunpoint was unimaginable in Concord, Massachusetts, in 1973. Denial, I would learn, is immensely seductive. It is irresistible for bystanders who want to get on with their lives. In the moment of terror, denial and dissociation are life-preserving for the victim. But over the long term, denial can be dangerous. In this case, the denial of our community resulted in many additional child rapes—at least forty-four—and the suicide of at least one of the victims.

In the police files, I would rediscover not only details of what occurred during the rape itself but also my own unusual reaction to the crime, and the unusual reaction of my family. There was a reason why I was drawn to spying on violent men. There was a reason why I was so good at it: I had done it my whole life, as a way to tame them, and to tame my own terror, the terror I didn't feel.

The book traces the investigation that the police undertook to solve this long-unsolved crime, and my own investigation into how the perpetrator came to be a rapist of children. It also traces my inquiry into the reasons I had acquired the symptoms of post-traumatic stress disorder. This inquiry into my own history was far more frightening than interviewing terrorists. It would have been impossible to do this, to enter the eye of sheer terror, if I didn't have a lot of help—a therapist who is an expert on trauma and dissociated states, and a partner who encouraged me to investigate my rape and who was willing to hold my hand while I did.

Jumping into the Sea

Chet keeps trying to get me to swim with him. I don't like to swim, I tell him. I keep telling him that. He took my son, Evan, and me kayaking. Evan and I beached our kayaks on the hot sand. Chet and Evan swam, and I watched, delighting in their delight, feeling a vicarious thrill of pleasure, the way you feel watching a puppy running in the wind. I pulled the kayaks farther in to safety, the stones sharp and burning under my bare feet. My skin was hot. I felt a premonition of ache in my temples; I hadn't drunk enough liquid. I ventured a few careful steps into the sea. The water was cooling to my feet. I wanted to want to swim. My son and my friend, who was not yet my lover, looked soothed and cooled, while I felt parched and oppressed.

I am not afraid of swimming. I just have no desire. And yet, the sound of lapping water alarms me. It reminds me of being on my grandfather's boat. Why should that alarm me?

Grandpa loved to take us out on his yacht, into Long Island Sound. Sometimes we went for long trips, spending our nights moored in the slips that dot the Sound. There was that soothing sound at night, the water lapping against the boat. I remember the comfort of the wind on my skin—a reminder that there was a world beyond the oppressive world of yachtsmen at play. It seemed to me that this yachtsmen's world was unsavory—that there were too many people there, that they were unserious, too loud. I wanted to escape into the world beyond the oppressive luxury of these yachts, bobbing corruptly in their slips, but I wasn't fully persuaded that it existed. Perhaps I had imagined this cleaner, purer world.

We would often stop en route to our nightly destinations and lower the anchor in the middle of the Sound. If you were quiet, there was that lapping sound. My grandmother would make us lunch, and Grandpa would take us swimming. It was exciting to swim in the middle of the bay. But it was also alarming: there was no beach, no clear horizon, no way to anchor yourself in a specific location. There were jellyfish floating near the top of the sea sometimes, and the thought of sharks below. There was also Grandpa's body.

My sister and I had vivid imaginations. Suppose a shark bit our legs, and tried to pull us down into the deeps, where he might dine on us, undisturbed? Grandpa was a strong swimmer. He might be able to save us if a shark bit our legs. And he could, in principle at least, sew our heart cavities shut if a shark tried to bite out our hearts. He was a doctor. He knew how to remove warts, slice through bunions, irradiate tumors—even imaginary ones. He would know how to sew up shark wounds with a needle and thread.

I know a surprising amount about my grandfather's body. When I think of my dead grandfather now, I sense not his spirit but his corporeal self—the pasty, humid white flesh; the irritating hairs. He was old and fat and pallid. But he was not a sweet old man: he was strong as a shark. I associate certain odors with him—the smell of decaying teeth that infected his breath, a sour scent that clung to his undershirts even when they were freshly washed, the smell of the bowel movements that his body strained to release, making the small bathroom even more claustrophobic. It was even worse if he sprayed Lysol in the bathroom to purify the air. The combination of the two scents—the attempt to cover up impurity—made one dizzy, made one want to throw up. To me these are the scents of an old man, the scents of perverse thoughts. My grandfather's body was a fountain of shame and disgust. But it was also a source of comfort. He was strong and able enough to protect us, I thought.

When I had croup as a small child, he took me into the shower with him. I remember his holding me to his naked flesh, a flash of electricity in me, a mixture of comfort and discomfort. Was this comfort an acceptable substitute for my mother? Was she dead by then, or just too sick to hold me, too sick to intervene? And later, too, I don't know when, perhaps when I was eight or nine—another shower, another encounter with naked flesh, which by that age I was old enough to recognize as repulsive.

I had a recurring dream at that age and into elementary school. A sickeningly soft white slug that came to me in my sleep. It is almost impossible to bear the memory of that dream. Even now, a nauseating bile rises to my throat, and even now, I feel locked in a prison of fear and fury. I cannot get out, I cannot breathe; how can it be that there is no one who will help me?

In fact, there was no one to help me. My father had disappeared into a life of work and love affairs, unable to bear the pain of my mother's death. Although he was living with us at our grandparents' house, I rarely saw him, and thought he had moved away.

I was certain that the dream slug that came to me in my sleep had no right to be there, that he had no right to speak to me, that he was breaking the rules. I wasn't sure what rule he was breaking, but I had an intuition that slugs were not permitted to taunt little girls at night. Still, I was incapable of defending myself, rendered speechless by uncertainty. And there were no grown-ups around in this recurrent dream, or in my waking life, to clarify the situation—to punish the rule-breaker, or, if need be, the false accuser. The dream slug stood up, looking like the Pillsbury Doughboy. He would tell me, "You must learn to play along. You are a bad girl. You will never get away from here, and you will never succeed."

I loved to swim in my teenage years, as long as it wasn't the Long Island Sound, which I saw as polluted. I liked the cold, pure waters of northern Maine and Canada. I liked to bathe in mountain streams, the water so cold it knocks your breath away, making you new, washing away shame. But later, the thought of slimy things made swimming unbearable for me. Swimming pools became chlorinated toilets. The sea had snails and slugs and snakes in it. It wasn't that I was afraid; I just had no desire.

A month ago, Chet tried again to get me to swim. I don't like to swim, I reminded him. We were in Vermont. We were walking by a stream. It was clean, with deep pools and occasional rapids. If you jumped in, the current carried you to the other side, where there was sand to rest. Somehow, I couldn't resist swimming this time, and as a result, I rediscovered exhilaration. But I also felt something like fear.

Later, he took me to the seaside, to a tidal pool that you could swim in, with narrow openings to the sea on either side. The pool was pristine—constantly washed by the shifting tides. And in that pool were tiny jellyfish. They didn't sting, but they slipped against your body like a hundred flaccid penises. No pain other than the agony of disgust. Eventually, it was more than I could bear, the terror of surreally soft flesh sliding against my unprotected skin. I begged him to hold me, to carry me out of that water, and he did. And he also promised to take me back to the sea so that I could vanquish that sea slug, my terrors, at last.

chapter one

Rape

I know that I was raped. But here is the odd thing. If my sister had not been raped, too, if she didn't remember—if I didn't have this police report right in front of me on my desk—I might doubt that the rape occurred. The memory feels a bit like a dream. It has hazy edges. Are there aspects of what I think I recall that I might have made up?

In the fall of 2006, I got a call from the police. Lt. Paul Macone, deputy chief of the police department in Concord, Massachusetts, called to tell me he wanted to reopen our rape case. "I need to know if you have any objection. And I will need your help," he said. The rape occurred in 1973.

Lt. Macone and I grew up in the same town, Concord, Massachusetts, considered by many to be the birthplace of our nation. It is the site of the "shot heard round the world," Ralph Waldo Emerson's phrase for the first shot in the first battle of the American

Revolution, which took place on the Old North Bridge on April 19, 1775. The town is frequently flooded with tourists, who come to see the pretty, historic village and the homes of Bronson and Louisa May Alcott, Ralph Waldo Emerson, Henry David Thoreau, and Nathaniel Hawthorne, who lived and worked there. It is still a small town, with small-town crimes. The *Concord Journal* still reports accidents involving sheep and cows.

Although we didn't know each other, Lt. Macone and I overlapped in high school. We never met back then; he was a "motor head," as he puts it, obsessed with cars, and I ran with an artier crowd. But I knew the name—everyone in town did—because of Macone's Sporting Goods. Everyone bought sports equipment there. It was a fixture in our town. It's where I bought my bike, the bike I was riding home from ballet class on the day I was raped.

I had recently requested the complete file. I wanted to understand what happened to me on the day that my sister and I were raped. I had an idea that by reading the file, by seeing the crime reports in black and white on a page, I could restore a kind of order in my mind. If I could just connect fact with feeling, the fuzziness in my head would be reduced, or so I hoped.

The file had to be redacted. Lt. Macone had to read the entire file in order to black out the names of suspects and other victims.

He told me, "I read that file from cover to cover. And I realized that the rapist might still be out there. There were other rapes. The same gun, the same MO—what if the rapist were still on the street? Other children could be at risk. I was worried about what might happen if the rapist were still at large."

Lt. Macone brought the case to his boss, the chief of police. "I've been a cop so long, we can't help trying to solve crimes. Twenty-nine years on the force. And I thought this crime was highly solvable. You'd have to be brain-dead not to see that.

There was the unusual MO. The fact that you and your sister both saw the perp's face before he put on his mask. There was a description of the gun, of the perp's clothing. The fact that he spent so much time in your house, early in the evening, when all the neighbors were home. To say that it piqued my curiosity is an understatement.

"It was clear to us what had to be done," Lt. Macone said. "We had to try to solve the crime. But we knew we would need your help. I needed to ask questions that could be quite personal. I didn't know if it could put you over the edge. . . . I didn't want to be a party to revictimizing someone. You seemed like you wanted to know . . . you seemed sincere. . . . But I needed your help to go forward. Not every rape victim wants to revisit the crime," he said.

But I was willing.

In the police station in Concord, files are kept in a locked room on the second floor. To get there, someone has to release the lock on the elevator, and someone has to let you into the locked room. Every time I'm near that room, I feel cold. There is a single file cabinet containing sexual assaults. One of those old-fashioned gunmetal gray cabinets. I hate those. The same kind my physician grandfather kept his X-rays in. The Concord PD keeps the records of certain crimes for a specified number of years, and the records of other crimes forever. The "forever" crimes include violent sexual assaults.

In the forever-crimes file cabinet, a couple inches back, was another fat file, another cold case. There had been a rape, two years before ours, right around the corner from where our rape occurred. "I was astounded," Lt. Macone told me. "There were two victims, just like your case. The same kind of gun. The perp sounded remarkably similar. Unless I was missing something, this was clearly the same guy.

"We get a lot of record requests," he said. "But this one was very unusual," he said. "This was the most serious crime I've seen in my twenty-nine years on the force."

"What was so unusual about the crime?" I asked. I had always heard that rape was a relatively common occurrence, so I was surprised.

"Everything," he said. "There are very few stranger rapes in Concord. We get indecent A and B's [assault and batteries], but very few crimes involving firearms. And very few cases involving home invasions, especially with kids involved. Every horrible factor you could have was there," he said.

It was almost impossible for him to imagine, he said, that a crime like this could occur in his hometown. It is almost impossible for me to imagine that Lt. Macone and I grew up together. Lt. Macone grew up in a safe town, a town where the few crimes that occurred would certainly be solved. I didn't.

Lt. Macone saw that the detectives who worked on the case in 1973 did not take the crime seriously, in part because they did not believe my sister and me. They had trouble believing that the rapist was a stranger to us. Rapes like the one we described simply did not occur in our town, or so they believed. Denial and disbelief were the easier course. The detectives left notes such as the following: "I told Mr. Stern that I feel the girls were holding something back from us," and "I was sure that this person may have been there longer in the house [than the girls reported to us]." On the back of one of the statements taken from me several days after the rape, I found a detective's handwritten note recording my reaction. "She says she sees she has a 'skill'—becoming very stern and hard." This was a clue, I would later discover, that I had already been traumatized. But how?

In notes from February 13, 1974, four months after the crime, I see: "Personal visit. Spoke to Mr. Stern. He states nothing new to add. He feels that both girls seem to have forgotten it." The police took my father's statement as permission to cease investigating the crime, and the rapist was not found.

I see in the files given to me by the Concord police a laboratory report from the Commonwealth of Massachusetts. A doctor examined us, I remember, in the hospital. It was dark by the time we got there, and the hospital was empty. I remember bright lights shining on my body and in my eyes, like a crime scene in a dream. In the dream, someone had accidentally transferred my soul to someone else's body for the evening. My sister Sara was in the hospital, too. But she must have been down the hall because I couldn't hear her voice. I felt alone. It was a time of night when we should have been in bed at our father's house, the beds my father had made for us. I wanted to say, "There has been an error here. This is me, Jessica. I am not a grown-up! I should not be alone here. I am not a criminal!" But I was stuck in the dream. I couldn't escape. I heard that buzzing sound of fluorescent lights. The sensation of cold when the speculum pierced, once again, the flesh that felt torn.

Would I ever heal? No, I would not. I would become someone else.

The laboratory reports indicate that sperm were detected in my underpants, my leotard, and my jeans. I do not recall whether they found sperm on my body, but they must have. How much of what I think I recall is correct? How much of it is a result of memory distorted by the Valium and sleeping pills the doctors administered to us afterward, or the chemicals my body produced, the chemicals that even now are distorting my features as I read the file, making me hard and tough?

How much of what we think of as an admirable response to trauma—the "stiff upper lip"—is actually disassociation, the mind's attempt to protect us from experiences that are too painful to digest? I can recall the facts, at least some of them. But I don't feel very much. At least, the feelings I have are not kind. They are not sympathetic toward my fifteen-year-old self. It happened. It happens to a lot of women. I survived. Most women do. I am "strong," but in those moments of strength, I don't feel.

I will admit that I am very afraid of one thing. Not just afraid. Ashamed. I am afraid that I am incapable of love.

This is not the first time I have read my police file. In 1994, when one of my neighbors thought she might have been sexually abused, it occurred to me that our rapist might have raped her, too. I asked the police for my then-twenty-year-old file, and they sent me part of it, including a barely audible audiotape. Even now, I haven't found the courage to listen to it. There was nothing in the file to help my neighbor, and nothing in the file that helped me, at least not then.

The first time I saw the file, it was incomplete. But now I have the whole thing. Everything is here—the things we had to tell the police, the things the police wrote down that they thought about us at the time. Only the names of suspects are blacked out.

I have trouble reading it. The copy is bad, but that is really just an excuse. I have trouble focusing my mind, as if my brain were underwater. I have an urge to put it away. I force myself to read. Moments later I want to stop again. Again, I force myself to read.

A wave of loneliness washes over me as I read. I felt so alone as a child. Our mother had died when I was three years old and my sister was two. We lived with our grandparents while my mother was dying, and stayed with them, after that, for a year. We moved out when my father remarried, a little over a year after our mother's death. He married Lisa, the older sister of my best friend in nursery school. The marriage had been more or less arranged by my best friend's mother and my grandmother. It was as if our father had married Mary Poppins. Lisa was young and bright and impossibly beautiful. I can still recall the comfort of resting my cheek against the smooth brown skin of her forearm. But it was not a good marriage. It lasted only six years. When I was twelve, our father married for the third time.

Our father was in Norway with his new wife at the time we were raped. I remember this distinctly. That is why a babysitter was staying with us. She was supposed to be ferrying us around. The evening of the rape we had gone to Lisa's house, as we did, once a week, after ballet. But Lisa went out to dinner that night, taking our half sisters with her. My sister and I stayed behind. We had homework. We asked the babysitter to pick us up. She was busy, she told us. I remember this. And I remember what I see here in the notes taken down by the police at the time: the babysitter didn't believe us when we called again a couple hours later, immediately after the rape, telling her that now she really needed to come get us. It was too hard to believe, in 1973, that girls could be raped in Concord, Massachusetts.

Our father was visiting the Trondheim Institute of Technology. I know because he told me this. And I remember. He was establishing what he considered to be a very important relationship for his laboratory. They would cooperate on radar technologies. But here is something I learn upon reading the file for the first time, something that, amazingly, I did not recall. When our family physician called our father to tell him his two girls had been raped at gunpoint, he did not curtail his trip. He did not come home to us right away. How is it that I keep forgetting this fact?

My father is a remarkable man. He is immensely strong. You can see his strength in the line of his jaw, his erect posture. If you met him, you would probably come to love and admire him. Nearly everyone does. You would sense an old-fashioned kind of integrity in him, an integrity based, in part, on the power of will.

I am proud of my father. He is like a precious jewel that has been handed down in my family. But he has sharp edges. Sharp edges, yes—but there is another side of him. He is wonderful with children. He lets them push themselves. He knows my son is agile and strong, so he let him climb on the roof of his garage from the time he was six years old. He would not let other, less strong children do that, my father tells me, and I believe him. I will admit that I feel slightly frightened when I see my son climbing on this roof, but I still my fear. I do not want to overprotect my son.

I suppose my father must have felt me to be tough, like my son. I know my father loves me very much, but sometimes the jewel-hard side is the one he displays, perhaps to protect the more tender side. I believe that my father is unbearably tender inside, but my belief is based more on faith than evidence.

My father invented many things, mostly related to surface-wave devices. He invented a flat-panel display. He and two other physicists invented X-ray lithography, a technology that makes it possible to lay X-ray-absorbing metals, such as gold or tungsten, onto a chip made of diamond or silicon. X-rays are what killed my mother. If only X-ray-absorbing metals had been placed on my mother's body when my grandfather irradiated her. But in those days the dangers of X-rays weren't fully understood, and my mother died.

There are items in the file that I find impossible to believe. For example, I told the police that someone—perhaps the rapist—called me at my father's house a few days after the rape, calling himself "Kevin Armadillo," identifying himself as "the one who fucked you last night." How could the rapist know to find me at my father's house? He did not rape us there, but at our first stepmother's home. How would he have known that we lived with our father, and were only visiting our first stepmother? I am mortified. I don't recall any such thing. Was I in a state of hysteria, making things up in order to get attention?

I also reported to the police that someone had left a handwritten poem in my bedroom, soon after the rape. I see in the file that the police questioned my father. He, too, had a hard time believing me. How would a person get a poem into the house without being noticed? How would he know when it would be safe to stop by our house, walk up to my bedroom? How could he know when no one would be home? Neither my father nor the police believed this strange story then, and I don't believe it now. I see what I told the police and feel ashamed. Why would I lie to the police?

But in the back of the file, I find the poem. Here it is:

Out

In

Out your window (or mine)

Flies an aged child

Who often lives from only a few seconds

But other times may become great and mighty

Strive with this flier

Its course is true to its laws and design

And note carefully: once it passes you,

It is gone; yet it has cooled and refreshed

This gentle wise flier is merely a zephyr

—a cool refreshing breeze

—a passing encounter

Touch and reflect off of and upon what

you touch
Leave behind the past
Steer carefully only for the immediate future
Be here now.
Trauma of the past must be understood as
Not being here now or it becomes trauma
of the present.

How could I have forgotten this poem? I do not know what to make of it.

I do remember the fact that I was raped. I can force myself to recall certain details. The fact that we thought he was joking when he told us he had a gun. We thought he was joking until he threatened to kill us. The fact, missing, for some reason, from the written record, that he asked us to show him where the knives were kept.

The police asked my sister and me to write down what had occurred. We did this, the report states, between 11:30 PM and 1:45 AM. Those words are before me now. I read the words I know I wrote, in a penmanship I barely recognize. My notes from 1973 are written in italics below.

—sitting doing homework
—man walked in

The penmanship looks alien. I don't recognize the persona captured in these words, the writing slanting backward, the letters round, fat. Was I ever this feminine?

I try to imagine "man walked in." I feel a kind of chemical strength. Not fear, not sadness, but a chemical agitation.

—showed us gun, don't scream

Reading these words, I feel a jarring in me, something quite hard and harsh forming in my veins, as if my blood were forming shards. This is a familiar feeling. I become a soldier if I am truly threatened. If the plane goes down, you want me at the controls.

Here is what I think now, reading what I wrote down for the police at age fifteen, right after I was raped. I was a good girl. Always a good girl, even when I was bad. I did my homework.

If I can only be good enough, someone will eventually notice that I am trying so hard, exhausting myself with my effort to be good. This is true even today.

I knew what good means. Good means never revealing fear. Good means not complaining about things that can't be changed, like the presence of a strange gunman in the house instructing me not to scream. Of course I won't scream, I didn't scream then, I won't scream now, certainly not out of fear or the thought of my own pain. I was not one of those hysterical girls who flinches or screams in the face of "man walked in" or "showed us gun." I knew that if I were bad, if I revealed my terror, he would kill me. I already knew how to absorb fear into my body—my own and others'—to project a state of utter calm and courage.

—said look down

Of course I looked down, not at his face. I understood the simple bargain: no looking, no flinching, no sudden movements, stay alive. Looking where he said to look.

What terror would I have seen in his face? I would have drowned in a sea of fear.

—he'd only be 5 min or 10, wouldn't hurt us
—was anyone home? When would they be home? Be quiet. Would kill us if we uttered a word.

Well, that is easy, only five or ten. If anyone else were home, they might not have known how to handle this man. But I did. Be quiet, he said. If you say anything, he said, I will kill you.

I was quiet.

—made us go upstairs looking down, quiet dog

My sister remembers this expedition, the three of us walking single file, up the narrow flight of stairs, his gun to our backs. "I thought we were being marched to our execution," she recalls. "I was trying to telegraph to you to be quiet. My biggest fear was that you wouldn't comply and that you would get killed." Why was my sister afraid that I would be the one killed?

In any case, I did comply. I floated into docility.

—close shades

The shades hiding our shame.

—take off pants. Asked if we'd still be clothed

The pants covering our shameful spots. The vulnerable spots now exposed. Would we still be clothed? We were wearing leotards. But I know the answer to that question now: I would never be clothed again.

—take off leotards.

We had ridden our bikes home from ballet class, jeans over our leotards. We were unspoiled and tough, unlike other girls, the kind whose parents might have driven them across town, the kind who might have screamed.

But now skin is revealed, legs exposed to cold air. Still, I did not scream.

—facedown on bed

I recall thinking that if I did what he said, we would stay alive. Don't scream, don't protest. I cannot recall the sensation of facedown on bed. Did the blanket scratch the cheeks and mouth, the mouth that would be good, that would not scream? Did the blankets comfort, did they suffocate?

—made us brush each other's hair
—made us try on little sister's dresses
—too small
—made us put on stepmother's dresses
—made us take off dresses
—told us to lie facedown
—made us sit up

How did those dresses feel on the skin that no longer seemed like mine, the "I" that I no longer know? Did I know that soon after putting on a dress, I would have to take it off?

I do recall "man walked in": I can see a kind of apparition in my mind's eye. I do recall the threat to kill us if we spoke. But now I am lost. My mind cannot focus. An apparition of cold flits across my heart but is gone so soon I wonder if I imagined it. I am annoyed with this little girl whom I'm struggling to hold in my mind's eye, who wants me to understand how she suffered. You will be fine, I want to tell her. I feel anger at her, even more than "man walked in." I do not want to hear about her fear or her pain. It wasn't that bad.

—stroke and lick penis
—said he put gun down
—said he could reach for it at any time

Now I begin to feel something new. A foggy nausea takes hold, leaving no room for thoughts or action. Why didn't I bite hard? Would it have been worth it to hurt this man, even if he killed me? Did I have the strength in my jaw to bite? I think not. I was in a sea of nothingness.

—sit with legs spread

Who spread those legs? How vulnerable I feel, thinking of this girl, her legs spread wide, exposed to this "man walked in," exposed to an evil cold.

—asked us what we called vaginas.
—we said crotches.
—he put his finger up me
—had we heard of cunt?

Yes.

—entered me while I was sitting.
—told me it didn't hurt—he was sterile and clean. Two times.
—I said it hurt
—he said it didn't.

I do remember the hurt, as if someone had inserted a gun made of granite that scraped my flesh raw, at first scratching, then tearing, then scraping the flesh off bone, leaving the bone sterilized by pain.

I am hollow and sterilized now. Not long after the rape, I lost my ability to urinate. I had to be catheterized, and later hospitalized. I began to walk with crotch held back to prevent intruders, muscles so tight I have to will myself to urinate, sometimes even now.

Now he turned his attention to my little sister.

—tried to rape my sister Sara

—I told him not to, please.

How did I find the strength to talk? I was spellbound by the potential for death contained in that gun, entranced into a statuelike calm. An animal intelligence had taken over where an "I" once held at least the illusion of dominion, where thoughts and action were once connected. The "I" was lured away into a space of infinite white, a space of no feeling other than calm, far from the human world, entranced into leaving its normal home—my body—by this man's insistence that he would kill me if I spoke. The animal mind that took over when the "I" had gone ordered the body that remained behind to be passive, silent, and calm, knowing, in its animal way, that compliance was required to keep the body alive.

But now there was something more important than sustaining the life of that body, something that knocked that shameless and shameful animal-mind back to its rightful place, a place I know nothing of, that I want to know nothing of, in this life. My little sister's pain pulled me out of my trance, and an "I" returned, determined to protect her; but I don't believe that the "I" that came back was the same "I" that my body had housed before. The new self that emerged was like a baby, having never been exposed to the world. The world felt new to me, baby that I was, more penetrated by sound. The sounds that had once thrilled me with feeling now grated on my ear. I had been playing Bach's third

French Suite, Debussy's "Engulfed Cathedral," and Beethoven's *Pathétique* before the hour the rapist spent playing with our lives, perhaps, so we thought, planning to kill us when he was finished with us. Where I once heard—in my mind's ear—a cathedral rising from a Turner-like mist, I now heard scrapes and moans rising from the piano keys under my fingers when I played.

I have trouble forcing my eye back to the page where I wrote the actions performed on me and by me during that very long hour.

—tried to rape my sister Sara
—I told him not to—please
—stood up. Told me to stand. Picked me up. Entered me telling me to wrap my legs around him.

Was he just a broken boy, needing someone to wrap her legs around him? This thought nauseates me again. A broken boy, stabbing and piercing a broken girl, leaving her shattered, as he was shattered. Why did I perceive him as broken even then, before I knew anything about him, before I knew anything about violence?

—faces down on bed
—told us it would make us angry

I remember what happened next: the click of his gun. I thought he was cocking it, preparing to kill. I was calm again, entranced into complying with his murderous plans.

Here is what shames me to the core: I thought he was going to kill me, but I did not fight him. I was hypnotized into passivity. I had no strength to run, and anyway I did not like the idea of being shot from behind. It seemed easier just to wait until the murder was done with.

There was no sex in that room. No love. But there was a seduction. The seduction of death.

Would he kill just one of us, making the other angry? Would he kill both of us, imagining that we would be angry at him in heaven, after our deaths? Why didn't I get up? Why remain "facedown on bed"? Why did I not rise up, Medusa-like, eyes flashing, the snakes in my hair ready to strike him dead? Why did I not overpower the puny little man, smack that gun out of his paltry, worm-white fingers? I was strong then, probably stronger than he, certainly very strong now.

And then he explained to us that the gun was not real, it was a cap gun.

—it was a cap gun.

I wrote dutifully, always dutiful. After complying with the rapist I complied with the police. Was this the most embarrassing part—that I had been entranced by the thought of a gun? That my fear, unfelt even as it was, had hypnotized me into complying with a person, if he was indeed a person and not an apparition, wielding a child's toy, not an instrument of death?

—he said don't call police. I promised I wouldn't.

—it would make us in more trouble.

—he left. We heard car.

I remember this part, too. I told Sara he was right; we shouldn't call the police. Somehow, even then, I felt him as a victim. I told her that they would put him in prison, that prison would not reform him: it would make him worse.

After he left, I saw in my mind's eye the image of a broken man, more broken still by the violence he would encounter behind bars, emerging as a true monster, a rapist who would actually kill his victims rather than leave them only half dead. It may be, I know now, that this intuition was correct. Was this a kind of Stockholm syndrome? Does it happen that fast, in the space of an hour?

Sara was more afraid than I, but also more alive. She retained a human-like strength in her arms and hands and mind that I now lacked. She insisted. She picked up the phone. No dial tone. He had cut the wires. He had cut the wires in the basement, a big puzzle. How would a rapist have time to cut the phone wires or know where to find them in the dark, dank basement? How long had he been in the house? How long had he been plotting this crime?

"He kept saying he wouldn't hurt us. He kept saying to listen, to be quiet."

I was quiet. I listened.

I'm still listening now. I hear a rush, in my mind's inner ear, of insistence. A kind of aural premonition, but a kind of premonition that goes both backward and forward, the soundless protest of all the raped, shamed, and silenced women from the beginning to the end of time. "He hurt you, he altered you forever," the chorus soundlessly insists, grating on my inner ear—the ear that wishes not to be reminded of feeling. I respond to that chorus: Hurt is not the right word for what that man—if he was a man and not an apparition—did to me. I feel a void. Something got cut out of me in that hour—my capacity for pain and fear were removed. The operation to remove those organs is indeed painful, as you might guess. But the surgeon cauterizes the cut, and feeling is dulled, even at the points where the surgeon's knife entered the once-tender flesh. There is no more tender flesh. It's quite liberating to have feeling removed, the fear and pain of life now dulled.

Nabokov once said, "Life is pain." Buddhists, too, believe that to live is to crave and to crave is to feel pain. To live in this world involves pain. Had I not been catapulted, in that one hour, halfway to death, and therefore closer to enlightenment? In death we no longer feel human cravings, no longer feel human pain. I was now halfway there.

Later, of course, I would come to reject this understanding of what happened to me that day. Yes, I was partly released from the pain of being alive. But my spirit had traveled, not toward the infinite divinity of enlightenment, but toward the infinite nothingness of indifference. A soft blanket of numbness descended like snow from the heavens, obscuring and protecting me from terror. Instead of fear, I felt numb. Instead of sadness, I experienced a complete absence of hope. I would come to feel, in a very small way, the indifference of the Muselmänner, the Auschwitz slang for the prisoners who had lost all hope, who no longer fought to stay alive. A divine spark, the craving for life, had been extinguished in those prisoners, who were destined for death. Their hopelessness turned out to be a death sentence. The other prisoners knew, by the blankness in the eyes of the Muselmänner, that these prisoners—whom Primo Levi referred to as "the drowned"—would soon die. The still-living prisoners avoided "the drowned" like the plague, as if indifference were contagious.

They were right, of course. Indifference is a dangerous disease.

"He kept saying to listen, to be quiet."

I have listened and I have been quiet all my life.

But now I will speak.

At the very bottom of the page, in a penmanship slanting ever more backward, I finally focused on what the police really wanted to know, the appearance of the apparition that had visited itself upon me.

—he smelled

—brown wool over face

—shorts, white socks

And then more detail.

—bobby socks, sneakers

—he was skinny

—light brown hair on legs

—strong cologne

—concord accent

It must have been hard to report the evidence of my senses—it came last, as if it were too painful to record anything other than the facts that transpired.

In the margin at the end of my statement, the police officer wrote what must have been my words to him at the time: "Forcing myself, determined to get it out."

My sister's statement reported the same facts, but her style is quite different from mine. She wrote in whole sentences, rather than lists. She began with how the rapist looked—his height, his socks, his sneakers. She remembered some things that I seem to have forgotten. That he made us lie down on the rug in the living room. That he had the woolen mask in his hand when he arrived. That we did not believe him at first when he said that he had a gun. She observed that he acted, throughout, as if he were teaching us.

After the rape, I fell into a perilous numbness, but fortunately, my sister took charge. Sara was petrified, but also determined to get help. She had the thought of walking out of the house into the cool night and going to use a pay phone on the street. So we did. That phone, too, was broken. We seemed to have entered a new, separate world where there was no way to communicate with the people we once knew.

Once again, Sara came up with a plan. We went to Friendly's. There, finally, we found a phone that worked. We called the babysitter who was in charge of us while our father was away in Norway.

Yes, we had been visiting our first stepmother, Lisa, our father's ex-wife. We visited Lisa, and our half sisters, every Monday night after ballet. But on that particular Monday night, October 1, 1973, she went out to dinner with our half sisters, leaving us behind to do our homework in an unlocked house in a safe neighborhood in a safe town, a town filled with good girls, though we were especially good. I was a good girl—I always did my homework, even when I was bad.

chapter two

The Legacy of the Holocaust

The phone at Friendly's worked.

I still recall, or imagine I recall, the familiar, metallic scent of the coin; the silver coin sweating, or damp from someone's else's sweat sticking it to my palm; then the comforting click as the coin slipped into its familiar slot; the familiar sounds of the line connecting. Some things seemed right and real. The sheen of the dime. The phone. The waitresses' white aprons, taut across their bellies, their bubble-gum-pink dresses with short sleeves. Simple objects unfettered with emotions, objects I could see up close, were not entirely distorted. But I was aware, as if in my peripheral vision, that much of my world was now blurred, as if I were underwater or looking through the hazy shimmer of fluorescent lights in the dark night air. The shimmer had a sinister edge. Straight lines curved into asymptotes, but soft things seemed hard or dull. Friendly's felt unfriendly to me now, like the fluorescent-lit café of Hopper's *Nighthawks*. The waitresses—whom I had always thought of as tough but kind—now struck me as stupid and unaware. Their movements slowed to a crawl in my vision. They were unaware of the real world I had seen—a real world where gun-toting rapists could come into your safe suburban town and change your life forever.

The babysitter, who was staying with us in my father's absence mainly to ferry us around, was a high school teacher from a neighboring town. In that town, too, child rapists were beyond imagination. She, too, had been transformed by my experiences of the last hours into a person too slow and stupid to warrant notice. Nonetheless, when she saw our faces and our disheveled appearance, she finally believed us. She called our family doctor, who arranged for us to be examined at the local hospital.

My father learned that we had been raped when our family doctor called him in Europe. A long-distance call. I remember thinking that the call must have been expensive. But my father didn't come home to us right away. I have long known this, but I forget it, again and again. I have not been able to take in this information long enough to retain it in my memory, until now.

The doctors prescribed sleeping pills. But I didn't sleep. I lived in a haze, both night and day, until eventually the world slid back, more or less, into place. I no longer know what the world looked like before I was raped, but I know that afterward—for a week or so—I descended into a state of dizzying calm.

We stayed, until my father came home, with Lisa, our first stepmother, in the house where we had been raped. I remember that Lisa comforted us, that she took us in like baby birds that had fallen from the nest. I remember sensing that she was on our side. Other than her kindness, I remember little else from that period. I know that my father didn't come home to us right away, not because I remember the days that followed our rape, but because it says so in the police report I now have before me, and because my sister remembers. I can access very little from my memory of what transpired in the days after the rape. I don't remember trying to call our father on the phone. I imagine we would have thought such a call far too expensive to consider.

My father married Lisa when I was five years old, very soon after our mother died. She was only twenty, just out of college, still a child herself. Lisa's mother, Myra, was a wicked witch out of a fairy tale, not at all the sort of grandmother you would expect to encounter in real life. Myra pushed Lisa, her eldest daughter, down the stairs. She threatened Lisa's younger sister, Judy, who was my age and my very best friend at the time, with a kitchen knife. She would occasionally leave Judy, who was red-haired and "obstinate," on the median strip of a highway. Even now, writing these words, I want to gather up that redhead into my arms and keep her safe.

During my father's courtship of Lisa and in the first year of their marriage, we spent a lot of time with Myra. Every night before our bath, Myra would line the three of us up—my sister, our aunt Judy, and me—to wipe our vaginas clean with a rough dry cloth. I still recall the feeling of the toilet where Myra made us sit. It was one of those fancy ones, with a wide mouth, the kind that a child might fall into if she didn't clutch the edge of the seat. Myra pushed hard, as if this part of our bodies required extra effort and strength to cleanse. I remember the dizzyingly bright lights of the room, the tiny white tiles on the floor. I am not sure why Myra felt the need to clean out our vaginas before our baths, but she did.

Our new mother, Lisa, desperately wanted to be a grown-up, and my father desperately wanted someone to help take care of his newly motherless girls. With the encouragement of my grandmother and Lisa's mother, my father married the charismatic Lisa, imagining that his problems were solved.

Soon Lisa had two children of her own. But Lisa's desire to mother my sister and me was erratic. If you pleased her, she made you feel like the daughter of a queen. If you disagreed with her, her affection waned. Six years later, Lisa and my father divorced. My sister and I remained with Lisa for nearly two years, while my father lived on his own. Children need their mother, he said when I asked him why he left us with Lisa during that period. But Lisa had not adopted us. When she decided to start a new family with a new, younger husband, she asked my father to take my sister and me back into his household. Still, we continued our regular visits with Lisa and our half sisters, and it was there, in the evening, that the rapist entered what had once been our home.

After the rape, Lisa rose to the occasion. Calamity brought out the best in her. She took us in and nursed us back to life, as best she could.

Upon reading part of the file in 1994, I learned that my father had informed the police that my sister and I "seemed to have forgotten it" four months after the rape. I also learned that the police had concluded that my sister and I were lying when we insisted that we did not know the perpetrator, and that my father did not defend us against this claim. Nor did he insist that the police continue to investigate the crime. There was no evidence in the file that Lisa had spoken to the police at any point after the rape itself.

In the spring of 1994, soon after seeing the file for the first time, I wrote my father a letter asking him why he had remained in Norway.

My father sent me an e-mail from his MIT e-mail account. His response was organized, with Q's and A's clearly specified.

He explained that he had been told that we were receiving medical and police attention. Lisa was taking care of us, and he was scheduled to return in three days in any case, he said. He was laying the groundwork, he said, for a decade-long collaboration between Lincoln Laboratory, where he worked, and similar labs in Norway and France.

He seemed to assume that once I knew his reasons for remaining in Norway, I would understand and approve of his sensible decision.

Despite this e-mail exchange, I managed to forget, for a second time, that my father did not come home right away, and I am shocked, for a third time, when I reread the file.

In spring 2007, I ask my father to read the description that I wrote for the police, at their insistence, immediately after the rape. I tell him that I would like to interview him for a book about my rape.

He comes to my apartment, glumly, determinedly, dutifully. He has done his homework—he has read the description of the rape and the draft of the previous chapter.

Within a few minutes of my father's arrival, my worldview begins to blur, as if the very erectness of his posture makes me question the correctness of my own. His right angles are at odds with mine. And yet, there is softness to his eyes. He has the posture of a soldier with the soft brown eyes of a deer.

I discipline myself to take note of his looks. So handsome, my father is. I am surprised, as I am every time I see him, that a seventy-nine-year-old man could be this muscled. He stands erect, dominating the room, even in his present dejected mood. He is wearing jeans, a no-iron shirt, an old Harris tweed sport coat, military posture. Everything faded but neat. Athlete's watch. Gray hair, beard, chiseled features. We make pleasantries.

He picks up the *Times*, as if he doesn't know that I know him well enough to know that he has already read it at first light.

I am having trouble with my tape recorder. I cannot get it to work, though it worked when I tested it just before my father's arrival.

He takes the machine out of my hands; he is the engineer, and I am the daughter.

"You see, you don't stick it in like that; you stick it in like this," he says. It is unclear to whom he is speaking—perhaps to me, perhaps to himself, perhaps to the machine. He pushes the tape back in, just the way it was before he took the machine out of my hands. He asks about the batteries. I tell him they are new. He points to a flashing red light. I bring a large box of new double As—the same box from which I took two batteries twenty minutes before. I tell myself that with his engineer's eye, he will choose more reliable ones.

I notice that a calm descends between us. We've done this many times before. My father, at ease in the physical world, delighting in fixing some *thing* for his daughter. And I am delighted to be fixed. I wait, becalmed by the familiar role, in the knowledge that my powerful father can make all things right. Children and dogs obey his orders, which are mostly issued in a calming baritone. If they don't obey the first time, they learn.

The material world becomes more orderly in my father's capable hands. When we were children and there were suddenly too many of us, he built beds in the laundry room. When we needed more rooms, he erected walls. He was still climbing Mount Washington into his eighties. When I moved into the apartment where I am writing these words, he measured the walls and drew up blueprints, wanting to ensure that my pictures were centered and straight. But I am living on the second floor of an old house. The floors slant rebelliously. Despite the careful measuring and drawing, the pictures did not submit to my father's will. Undaunted, my father started again, this time with a level. The pictures succumbed.

But this time, although he's tried all angles, the machine refuses to work. He diagnoses the problem: the mechanism isn't working properly. It's old, I say. A cheap machine.

I bring him a cup of tea, and we settle into the task at hand: I will type as fast as I can, which is very fast indeed, while my father answers my questions.

"What are you most proud of in your life?" I ask.

"It changes with age. Now I am proudest of my children because they are extensions of me into the future," he says.

"Do you believe they are a good extension of you into the future?"

"Yes," he says. "They exhibit qualities I admire. They are independent. They have taken possession of their lives. And what they do, by and large, are things I admire.

"It's the lives they're living that I'm proud of. Sara has realized herself. She wants to sing, and she sings. She is a successful mother, a successful wife, and a successful professional. It's the life she leads.

"It's the same with Arabelle. I admire that she found a vocation for herself. I admire that she has the perseverance and ability to fashion a life for herself and be successful at it.

"Same thing is true with Genevieve."

He has started with the second-eldest and ended with the youngest. I am the eldest. Of course I wonder if or how he will find a way to say that he is proud of me, even though I am presently his interrogator.

"Same thing with you," he says, at last. "You are your own person. And the most important role a parent has is to help children achieve that status."

"How did you do it?" I ask, genuinely curious, hoping I will learn something that will help me raise my son.

"I don't know. By letting children struggle rather than jumping in. Helping them only when necessary. Encouraging them to take responsibility for themselves as soon as they were able. To the extent that I could, helping them gain an education that prepared them for life."

I watch my father pausing to reflect. I can see that he is determined do his duty—to get this right. I take great comfort from my father's integrity, even though I know that his desire to stick to the truth might sometimes hurt me.

"This is too narrow," he corrects himself. "I'm proud of your music. I'm proud of your achievements. Actually, I'm not proud—I admire the way you play the piano; I admire your achievements. I admire your writing ability," he says, giving me the opening to talk about the description of my rape.

Even though he has given me an opening, it feels as if I am breaking through ice.

I broach the topic: "Were you surprised by what you read? Did you know what happened when we were raped?"

"No," he says, answering only my second question.

"I didn't know about it," he continues. "After reading your tract of your rape, I notice that you have not made up your mind to put that horrible experience behind you and move on to the future as I did. Usually a person who is preoccupied with the past becomes basically dysfunctional in life. But you're able to be preoccupied and yet to function very well. I'm proud of that."

Is this really a compliment? I wonder.

I tell him that I'm deliberately going back over this material because of the book I'm writing.

(Vaguely, I am aware of a familiar defensiveness in my voice that happens when my father expresses disapproval of me.)

I start out slowly, a technique I have learned from talking to terrorists for so many years. Never with small talk; that would feel too manipulative. But I skirt the topic we both want so much to avoid.

This time, however, it is not only my interviewee whom I want to put at ease, but also myself.

"When have you been afraid in your life?" I ask him.

"I was very much afraid from the time I was six until I was almost ten," he says. My father is prepared to be brutally honest, even, or perhaps especially, at his own expense. But he answers me with little feeling, as if fear were something he once read about in a newspaper, but never actually experienced in his own body.

"What was your first experience of fear?" I ask.

"I had heard about my uncle Solomon being carted off to a concentration camp and then dying within three months. Nineteen thirty-three. He was a person who sort of made jokes. He saw some men pulling up metal fencing, and he joked that they were doing it to use the metal for weapons. He made a joke about it. He was standing next to a workman. My mother's brother-in-law. And they carted him away.

"Shortly thereafter I was playing doctor with a neighbor's girl," my father continues. "The idea was to discover what her crotch looked like. We were just playing. Afterward the town gendarme said he'd had a complaint about me; that I was defiling an Aryan girl. He told me that if that ever happened again, my whole family and I would be sent to a concentration camp." All this my father says plainly, in a kind of monotone, as if dutifully reporting the details of an experiment that didn't go as planned.

I don't know what my father saw in my face. Perhaps I sucked in my breath.

"That scared me," he explains.

"Was your mother there when the gendarme approached

you?" I ask, shocked by this story, which I am hearing for the first time.

"I was alone," he says.

As I read these notes now, months later, this seems like a good summary of my sense of my father's life experience: "I was alone." It has nothing to do with my father's actual life. My father has always been surrounded by people. Seven siblings. Countless cousins, despite the reduction in their numbers by the Nazis. Three wives. Seven children. Eight grandchildren. Nonetheless, I am compelled, once more, by a recursive thought: I don't want my father to die alone. I want us to be able to sit together, talking or not talking, whatever he wants. I want for him what I want for myself, to know that he is loved for who he truly is, including the pain he has felt and the pain he imposed on others, without denial. Isn't that what we all long for?

"Did you tell your mother?" I ask.

"I don't think so. . . . I doubt it. I'm not certain. I suppose I assumed that you're not supposed to do things like that, play doctor with girls, especially when the girl's father is a Nazi."

My father is still speaking in the detached tone he reserves for distastefully emotional subjects. I try not to distract him with my own emotion. But I have never heard any of this before, and I am nearly paralyzed with terror.

"You thought you had done something really wrong?" I ask, quietly, as if trying not to frighten the little boy in this story.

"Well, maybe not terribly wrong, but something that endangered my family and me."

I don't know how, in my paralysis, I summoned up the courage to push the question further.

"So your first sexual experience was connected with danger—with the threat of death," I say.

Reading these notes now, I can no longer grasp that my father and I managed to allow these words actually to exist between us, to be uttered aloud, these terrible truths that were not to be captured by words.

"I'm not sure I phrased it that way to myself," he says. "We were little. I was curious."

"So you discovered that your curiosity could be dangerous," I say, bravely.

How alike my father and I are, I realize now. My curiosity compels me. My curiosity is dangerous, too. My curiosity propels me on, even in this interview, the most dangerous and difficult one of my life thus far.

"I thought I was being persecuted because I was Jewish," he says, sticking with the facts.

I have come to believe that my father's trick, his whole life, has been to convert the sensation of fear, which makes him feel ashamed, into dominance. When my father is afraid of something or some powerful emotion he fears he cannot control, he finds a way to dominate himself and others. This is why some men go to war, I think to myself, and then brush the thought out of my mind. Too pat.

There is an aphorism that comes unbidden to my mind like a mantra whenever something truly bad happens: When the going gets tough, the tough get going. This is my father's recipe for resilience. This philosophy has often served me well. It makes me tough. But sometimes, I am ashamed to say, it makes me cruel—to myself and to others. Sitting here now, reading this interview, I recall with a nauseating shame my cruelty to an aunt who was going through a terrible divorce. She had recently discovered that her husband had gambled away much of their life savings, and she felt utterly lost. Although she was trained as a lawyer, she had never really worked, and now she was on her own. "Get a job!" I told her, sensibly, as if channeling my father's eminently sensible approach to life. I apologized when I realized what I had done, but I'm not sure that she will ever be able to forgive me.

"Did you think that being Jewish was bad?" I ask.

"I was going to public school then, and the first year all the little boys were put into the Hitlerjugend shirts. I wanted to do the same thing—it looked very desirable. But I knew I couldn't join because I was Jewish."

My father has a stronger accent now, his syntax slightly awkward.

"The second year a Nazi functionary was put in place as the principal. And this new principal came into our classroom and made all the Jewish boys and girls stand up—"

"How many of you were there?" I ask, trying, like my father, to stick to the facts of the story. Trying not to feel, or anyway, not feeling.

"About two or three of us. And then he said that Jews were the source of all the problems in Germany—he said they were enemies of the people. He reiterated the caricatures of Jews," my father says.

"What were these caricatures?" I ask.

"Money grabbing. They will steal your belongings, they've been stealing from the country and individuals. . . . I didn't listen that closely. . . . I didn't believe it . . . because I knew my own family."

I think to myself, *moneygrubbing,* but don't correct my father.

"What did being Jewish mean to you when you were a kid?" I ask.

For the second time he doesn't answer my question, but continues with his story.

"And he then told all the boys in the class that it was patriotic to beat up all the Jewish boys and make them go away. . . . They chased me and caught me and whipped me with these flexible sticks."

"What do you mean by flexible sticks? Branches?" I ask. My heart is beating hard.

"They were used in Germany to whip—"

"You mean switches?" I ask.

"Yes, that is the term," he says.

I am not aware of feeling emotion, except that something as simple as finding the right word has brought us both relief.

"Every time school got out from that day forward, I had to make myself disappear so as not to get beaten up. I thought I was succeeding pretty well. But one day I met one boy who was one of my tormentors, and I attacked him and beat him up. Again the gendarme came around and said that if I laid a hand on another Aryan youth, my family and I would be sent to a concentration camp.

"The Nazis would have beaten me up and sent me off without discussion. But he was the town gendarme. He knew me, he knew my family. Nevertheless it scared me very much.

"When school was out, a person from the Joint Organization, a Jewish organization, traveled through Wetter [the town where my father lived] and was talking to the Jewish families, and when she came to our house, my mother described what had been happening to me at school. The lady asked to see my back. My back was so full of welts. She suggested a Jewish school in Marburg, and I should be sent away from home and sent to a Jewish camp in Frankfurt for the summer. And I did that."

"What was the camp like?" I ask.

"I don't remember the camp. I just existed there," he says.

I have never heard about this camp before. My father has not talked about his childhood in Germany very much. When I've asked, he has told me about having to leave his kitten behind. And also his cousin, who was eventually killed by Mengele.

There are unstated conditions for this conversation. I will evacuate myself of feeling so as not to distract my father. This is an unspoken pact we've had ever since I can remember, since my mother died. Usually I can follow the rules.

Four years ago, when I was forty-six years old and my sister forty-five, we finally worked up the courage to ask my father to describe his recollections of our mother—a topic that my sister and I understood, from the time we were children, was off-limits. There had been no photographs of our dead mother anywhere visible in the house. We didn't know, until the time of this conversation, well over forty years after our mother's death, that my father possessed a photo album of their wedding, which he had hidden in a bottom drawer. Nearly all of our mother's possessions—the wedding gifts she received from family members, even her jewelry and silver—had somehow disappeared, given to other family members who apparently did not understand what these things would mean to my sister and me. I do not possess a single object that belonged to my mother. There was so much feeling in the room that day—something we were so unaccustomed to—that my sister requested that we break and continue the following day. But my father warned us, "I will not discuss this topic again." It was too painful, he told us.

My father continues with his story. "The next fall I started going to school in Marburg. I was afraid we would be a target because now all the Jewish children were congregating, at the railroad station, on the train.

"Then the SS started appearing," he says, continuing carefully.

"They were considered particularly deadly. I remember walking from the railroad station. Every time I'd see an SS man, I'd try to walk around him. Whenever it was inescapable that I had to pass him, I'd just steel myself for the process, try not to call attention to myself in any way."

There is a dry hollowness between us now. The absence of feeling leaves me light-headed. I am slightly nauseated, a familiar sensation. It comes to me when there is some strong feeling unfelt.

But with all this discussion of my father's attempts to avoid drawing attention to himself, feeling periodically leaks into the room.

I'm better at interviewing terrorists than interviewing my father. It's too close to home.

"Is this the sort of thing you want to know?" he asks. I see that my father is embarrassed by having to go over this distasteful story. He is determined to do his duty, but we are perilously close to an indulgent sort of introspection, which my father deems "examining one's navel." He hates to call attention to himself by divulging more than is required.

But there is something more: I sense my father's unacknowledged shame, and I feel ashamed of myself. Like a naughty child searching through her parents' drawers, I have now seen something I was not supposed to see—my father's shame. It's not only that I have discovered something illicit, but my dangerous, prurient curiosity about my father's inner life is now on display, exposed to my father's disapproval. My father does not approve of my "examining my navel," but I, ever the difficult child, want to examine not only my own, but his, too.

"Tell me more about what it was like to feel afraid," I request. I need to know. I need to know what it *feels* like to be afraid.

"I felt afraid," he concedes, dutifully confessing the truth as he knows it, even when it is embarrassing or painful to himself or others. But he has not yet provided me the details about the sensation of fear that I want so much to hear.

"No," he says, pausing, and then more firmly, "I felt terror morning, noon, and night. The brownshirts would decide they needed to teach us a lesson. And then they would practice putting out fires by shooting streams of water at our house—their fire hoses shooting right at us. They never did break in and attack us, but that could happen at any time."

There is a shift in the room now. More feeling, and with more feeling, relief.

"They marched into the house one time with guns drawn," he adds.

"Who was there?" I ask. My skin feels prickly.

"My mother, Irmgard [my father's next older sister], and Anna Marie's mother [a neighbor]. They marched in. Irmgard was fourteen, and I was seven. They took my mother into the front room and shut the door.

"I was looking at Irmgard to try to figure out what I should do. What she did was collapse on the flagstones in the front hall and start screaming and kicking her heels. So I tried for a little while to do that, too, but it didn't seem sensible. So I stood there, aghast. They were in there about fifteen minutes. My mother told us that what they had demanded was that she sign receipts indicating they had paid her whatever debts were owed to my family.

"I didn't know what to do," he says. "So I tried yelling for a while. Not long. And then I just stood there. I thought, Should I go and help my mother? I wasn't certain I should. I was afraid to go help her. I didn't hear any screams or anything. The door was closed."

"How many of them were there?"

"Three or four," he says.

"Were you afraid they were going to kill her?" I ask.

"I would have heard a gunshot," my father says, sensibly.

He considers my question a bit further. "I was afraid they would kill us all," he adds.

"So you just stood in the front hallway, your ear to the door, listening? Hoping not to hear a gunshot?"

"I stood there. When they were done, they opened the door."

I repeat to myself, "When they were done."

Done with what?

"The same door, where you were standing?" I ask, trying to tether us both to the physical world with concrete details.

"Yes. I was in the front hallway. They had marched into the sitting room, where my father's desk was."

"What were their guns like? I ask.

"I'm not sure I even saw the guns," he says, as if thinking of this for the first time.

"Then how did you know that they had guns?" I ask.

"I always thought whenever the Nazis showed up, they had guns with them," he says.

"Do you think you saw the guns and then forgot them?"

"No, I don't think so." My father cannot imagine that he would forget something like this. I know that he could.

"I believe my mother told me afterward. I knew they had drawn guns, but I'm not sure how I knew."

"So the door opens. What happened after that?"

"I don't recall," he says.

"Did she come over and hug you?" I ask.

"I don't recall. I think she might have come out, and we hugged each other. She would have picked us up.

"I'm surmising," he explains.

"Where was your father?" I whisper.

"Apparently my father was in the fields at the time of the attack, when the brownshirts marched into our house. My mother was deathly afraid that my father would return with a scythe in his arms," he says.

Why was my grandmother afraid that her husband would go after the Nazis with a scythe? Could the Nazis have raped my grandmother?

Is my father telling me that he understands what happened to me, that he understands why I knew it was not sensible to be hysterical under the rapist's gun? Is he warning me that after this brief interlude he will go back behind the wall that shuts out the recognition of terror, the terrors of his past and the terrors of mine?

For now, in this very moment, we are allies, my father and I, in the war against terror and rape. In this very moment, we are together.

Now my father switches to the present.

"Except for the beating of the boys, I never suffered anything comparable to what you went through," he says, shocking me with a bolt of empathy. He seems to be considering my past experience in a new light. And with this return to the present, I sense new rules—he seems to be giving both of us permission to feel for the other, at least for the moment. We are no longer in an empathy-free zone.

"The beatings didn't bother me very much because in my family children were physically punished if they were bad. It wasn't as bad as it would be for a child today, for Evan, who has never been beaten," he says, referring to my son.

I am thinking now of when Evan was a toddler. Like many toddlers, when Evan got too excited, he would sometimes try to bite me. My father offered to cure him by biting Evan back.

"I prefer to be spanked than to be yelled at or made to feel guilty. It was less painful. I felt warm in my bottom and that was it," my father says.

"I was afraid at night, afraid during the day," he continues, without prompting from me.

There is a noticeable lessening of tension in the room.

"Did you have trouble sleeping?" I ask.

"But I didn't know I was terrorized," my father says, as if he hasn't heard my question. So he didn't feel his terror in the moment, either.

"When did you realize?" I ask.

"I still remember the feeling," he says. "As soon as the U.S. ship started pulling away from the dock in Hamburg, I felt a huge upwelling. Then I realized that my fear was leaving me—the ship was American, and American law applied there, although there were some Nazis on the ship. The USS *Washington* . . . a passenger liner . . . March of 1938 . . . so I was almost ten.

"I got very sick. I couldn't hold anything down. I'm not sure, maybe I got food poisoned. As soon as I recovered, I was voraciously hungry. And everyone else was sick, and I was getting to run around the ship. And getting to eat four or five meals a day. We didn't have much to eat in Germany—we had fake flour. I got to eat real food. I was small and thin for my age at the time."

It is hard for me to imagine this muscled man emaciated and starving.

"But I still had German habits," he says, by which he means, the habits of a German Jew, petrified of the Nazis. "If I saw Boy Scouts, I would go way out of my way to avoid them because they reminded me of Hitlerjugend. I was very passive and did everything I was told in school—I imbued any authorities with German power. I was afraid of them. Of the principal. That was with me until I got to be thirteen or fourteen. . . . . Even then I deferred to authority."

I hear again my father's words: "I didn't know I was terrorized." I wonder now, was he terrorized by the police in Concord, Massachusetts? So determined to get them off the case that he told them we had forgotten about the crime four months after it occurred?

"So only some of your fear left when you left Germany?"

"My anxiety left. But those other fears were bred into me. . . . It took a long time to realize they weren't founded on anything. Another fear is that I didn't ever want to stand out, so I feared for a long time huge visible success—I didn't know that I was smart. I didn't try to get prizes or anything like that because I was afraid I'd be exposed. In college. And even professionally for a while. It did have a lasting effect, living through that terror."

Is this why my father never praised us when we did well in school? I wonder now.

"This quality that you write about"—he changes the subject—"of not feeling afraid during a crisis. I know exactly what you mean. In a frightening situation I become very clearheaded. I assess the situation and try to maneuver within the constraints of the situation.

"When we had the car accident, I lost control of the car," he says, referring to a serious accident he and my stepmother suffered in 1994. "The car did not respond to the steering wheel, but I wasn't frightened. I tried to create a trajectory to cause the least damage. I didn't just throw my hands up.

"Same thing when I'm mountain climbing. If I get into trouble, I suppress everything that I'm concerned about. . . . I'm not afraid. It's gone."

"How does it feel to go into that state?" I ask.

I've only recently realized that there is something unusual about this capacity to slip into an altered state that makes me more efficient, even smarter. I am not sure how to feel about my father's admission that he can do the same thing.

"It doesn't feel good or bad. I feel competent. I feel I'm in control," he says.

"Afterward, do you feel burned out?" I ask.

"Physically exhausted," he says. "I feel tired and emotionally low."

"Did you ever ask yourself why you could do that?" I ask.

"No," he answers, simply, honestly.

We will return to this subject later.

chapter three

The Investigation

Once I read the complete file, I had to learn about the man who raped me. I needed to do this to tame him—but also to tame a wild, nameless feeling inside myself.

I have always been a spy. Whenever I sense pain that I don't understand, my own or others', I feel compelled to research the source. I become a detective.

This is embarrassing to admit, but I am insatiably curious about the half-known truths that motivate people's lives, often in ways they do not realize. If I met you today and sensed you have a secret—especially a secret you keep from yourself, especially a secret that might hurt someone—I would start trying to find the key from the moment I laid eyes on you. I might not even know that I was doing this. I might not want to do it, but I can't stop myself.

I have been spying on violent men for much of my life. Not

just men who have hurt me, but also men who have hurt others. I have traveled all over the world to talk to terrorists. I am compelled to understand men who hurt people, as if by understanding what motivates them, I can tame them; as if by taming them, I can make my world safe. But, perhaps for the first time, I am aware that my curiosity could make me sick. Sara tells me that she assumed, after the rape, that the next crime against us would be murder. She wasn't sure whether the rapist would come back to kill us, or a new perpetrator. But it seemed logical, she felt, that we would be killed. Is murder what I fear?

This time, however, I have help. I am spying together with the police. They want to find the rapist, too, for their own reasons. They want to get him off the street. Today we understand that violent pedophiles cannot be cured. They can be treated with medication, but even then, they must be kept away from children.

Paul Macone, my partner in espionage, wants to talk to me. He wants to give me some redacted files. There is more than one case that he thinks is similar to mine. Remarkably similar-sounding perp, he says. Very similar MO. Same grayish black gun with white grips. He asks me to come out to the station. He would like my help, he says.

Although Lt. Macone and I overlapped in school, I cannot connect with him in a personal way. He is good, in a way that I fear that I am not. Unspoiled. He knows things about me that I normally don't reveal. He knows about how the shadow of a long-ago terror debilitates me. I worry that he can sense an unhealthy obsession in me, and I feel shy in his presence.

I go to Concord regularly. It's not just that I grew up there. My father and his wife still live there, and my son visits them every week. I have driven past the police station hundreds, if not thousands, of times. I have a vague recollection that I've even been inside the building, but I don't remember anything. Was I ever arrested? I'm certain I was not. But shame shimmers at the edge of my consciousness, like a mirage I can't quite see. I discover that I cannot remember precisely where the station is.

I drive past the Louisa May Alcott house, which I visited several times as a child. I know the station is near here, but I seem to have gone too far. I drive back toward the center of town. I am terribly sleepy. There is the Scout House, a large eighteenth-century barn where I attended Girl Scout meetings, where I had to take dancing lessons. An image of white gloves floats into my mind. Did we really wear white gloves? I cannot recall; my brain is turned to mush.

After this bewildering incident, I don't even try to drive back there. Lt. Macone keeps me apprised of his progress by e-mail.

In October 2006 I open this e-mail.

Hi Jessica,

I had an interesting conversation with a detective in Weston, who has been around for years (and still working). What Weston does have is much, or all of the statewide intelligence on many suspects in assaults from the same time period. In a stroke of luck, they saved all of the paper intelligence so I will go retrieve it from them in the next couple of days.

Once I sift through what they have I will let you know if anything seems related.

I do not respond. Several days later he writes again. I open his message immediately, anxious to hear what he has to say.

Hi Jessica,

I have almost finished reading the intelligence we picked up at Weston PD. Nothing glaring is jumping out, however Lexington had two other assaults in the same general time period that I want to explore with them. In doing some process of elimination, I have a couple of questions. If you are not comfortable with any of the questions, please just let me know.

One report says there were obscene phone calls made to the house after the assault. Do you remember anything about this?

Let me know if I am being a pain. . . .

Paul

It takes me weeks to respond. I'm busy, I guess. And I also get sleepy. I want Lt. Macone to find my rapist because I want to interview him myself, but I am not able to provide much help. I am not aware of feeling afraid. But I don't feel like dwelling on the topic. I put the thought of my rapist and of Paul's continuing investigation aside. I will wait until the rapist is found, I tell myself, and think about it then.

It's not that I'm afraid of my rapist. It happened long ago, I tell myself. I'm grown up now. But I am afraid of the police station, which, for me, is a repository of shame. I prefer to stay away.

Once we find the rapist, I plan to talk to him. I plan to look him straight in the eye. That is as far as my planning goes.

I do not write out a series of questions, as I have often done in preparing to interview terrorists. I do not think about whether I will need to have a rifle, or perhaps a sword, or perhaps a posse. I always went unarmed to my interviews with terrorists. I always thought my very vulnerability was my best protection. But somehow, though I don't really think about it, I have an idea that this is different.

I do not plan that the interview will take place in a restaurant. I do not plan that it will be at Friendly's, an obvious choice. I do not think about how I will skewer him, not with a Friendly's butter knife, but by looking him straight in the eye.

If I make him see that I am not just his object, but also a subject, there will be an explosion in his brain. An electric signal will cascade: he will realize that he wronged the universe, and his brain will explode. Also his penis will fall off. I will leave him there, his brain on his plate. He will remain sitting upright, but the waitresses will realize that he's dead because the top of his head will have flopped neatly off and there will be wires sticking out. Broken phone wires, thick and tangled but cut through all the way, looking as though they'd been cut with shears rather than the power of thought. I will avoid stepping on his shriveled penis as I walk out the door. I will leave it there for the rats. I will not apologize to Friendly's for the mess: I cannot help it that electricity comes out of your body when you really look at someone. I may be a victim, but I'm a world-class perpetrator, too.

These not-plans, which I am describing to you now, take place in another dimension, a place I prefer not to dwell.

Although I am curious about what Lt. Macone is finding out, I don't drive out to the police station in Concord, the town where I was raped, the town where my parents still live, a twenty-minute drive from where my son and I live now. There is no point: I know I will get lost.

I e-mail Lt. Macone to tell him that I saw in the file that we had informed the police that I had received an obscene phone call shortly after the rape. But I don't remember the call, or much else that occurred during that period. Soon, however, Lt. Macone has more news.

Jessica,

Just now answered the phone from Natick PD. They found their file from their incidents!! It apparently has volumes of info. The detective is making me a complete copy of the file and I will head to Natick as soon as she calls back that it is done. I hope it is today. Let me know when you want to come

here to begin reading what I have found. I feel confident we are getting close. I will work around your schedule.
Paul

He will work around my schedule. He's gotten the message, apparently, that I am busy. It's not as if this investigation is my whole life.

Because I work on terrorism, I have good contacts in the FBI. After September 11, a number of agents who had worked on "ordinary" crime were recruited to work on counterterrorism cases, and a series of them have come to visit me over the years for what amounts to terrorism lessons.

I've never taken a fee for this work, although I've devoted many hours to it. The most intense period was in January–February 2002, when *Wall Street Journal* reporter Daniel Pearl was abducted by terrorists in Pakistan. I didn't mean to, but I all but ceased doing my own work during this period. I became obsessed with helping to save Daniel Pearl, as absurd as that might sound. I called all my contacts in Pakistan. It seemed to me very important to be in touch with the terrorists themselves, not just with Pakistani government officials or the Oxbridge-educated elite that Americans typically talk to in Pakistan. I had done the same sort of "stupid" things Pearl had done—meeting alone with terrorists, and admitting, when asked, that I was a Jew. But I had been lucky. My last trip to Pakistan to talk to members of bin Laden's International Islamic Front was in August 2001, and I survived, despite sharing Pearl's naïveté, as the media described it at the time. Pearl did his work after September 11, when the attitude toward Americans had shifted dramatically.

I identified with Pearl, of course. But there was more than that: I understand now that I was feeling, for the first time, the fear that I had not perceived when I had met with the same terrorist groups that Pearl met with. At the time of Pearl's abduction it didn't feel like fear, exactly. It felt like a premonition of evil.

Nine days after he was abducted, Daniel Pearl's captors slit his throat and beheaded him. I can almost feel that knife. A month later a film was released, showing the terrorists' horrific crime. The film was disseminated widely to recruit others to the terrorist cause. Despite the ultimate failure of our efforts to discover the captors in time to save Pearl's life, I received a letter of commendation from the director of the FBI for the futile help I provided during those weeks.

Like most people who have served in government, I have acquired a number of letters like this one, commemorating my service. I put these letters, and the other mementos I received, in a box. Then the box disappeared. I think it is somewhere in the basement of the house I once shared with my ex-husband.

But I have kept this letter from the FBI director. I never worked for the FBI, so that makes it more valuable. More like a stamp of approval from a higher power. More like a stamp telling me, "You're okay"; that this mysterious shame I feel, especially in the police station, is unwarranted, at least in the eyes of the FBI, at least for now.

One day it came into my head to ask one of the agents I knew to talk to Lt. Macone. It took some weeks for me to broach the topic. It was as if there were two persons living in my body—one, a tough, seemingly fearless person who traveled around the world talking to terrorists and whose knowledge turned out to be of interest to people fighting crime; and the other, a shame-filled victim. I had more or less left that victim and her unattractive, girlie feelings behind. I didn't think about her.

Victims—I somehow "knew" this to be a fact—are ineffectual, weak, and dishonest persons who drag society, or anyway their families, or anyway their fathers, down. I understood that terror and despair were contagious emotions, and that to indulge oneself in the feeling of terror was antisocial and possibly even immoral. I found a way to slice off the side of my self that felt endangered, and endangering, by the shameful feelings I could not stop myself from experiencing.

After the rape, according to a police officer's notes taken down on the back of his crime report, I reported that I had a "skill" in becoming "stern and hard" as a response to terror. I don't recall saying this to the police. But I do know that I understood, long before the rape, that to become stern and hard was a more "manly" approach to fear and despair, that it demonstrated a kind of good breeding, a kind of moral fiber. I wonder now why the officer took note of my words. Was he surprised to hear a terrorized fifteen-year-old girl speaking this way? And yet he does not seem to have asked me how I acquired this skill, or to what end.

Becoming stern and hard is so inbred in me that a more natural, "girlie" reaction to pain or fear takes an act of will. My sister, who was petrified to be alone for many years after the rape, was deeply ashamed of her fear. She, too, had taken in the notion that feeling afraid was somehow unseemly in our family. She explained to me later how she tried to finagle visits or phone calls with friends at times when she might otherwise be alone with the silence of an empty house. But she took great pains to disguise these efforts.

Talking to my contacts in the FBI about my rape—especially now that I was no longer able to speak about the crime as if it had happened to someone I didn't know—would bring back to life a shameful side of myself that I had tried to deaden, that I had pronounced not me. I'm not just talking about the different roles one plays in life—parent, spouse, professional—and the way different parts of one's character emerge in these different roles. I am talking about a part of the self that is almost wholly other, a victimized and now despised Siamese twin that survived despite my effort to extrude it.

Finding a way to be both persons was a terrifying prospect. It took all the willpower I could summon. It was worse than jumping into the sea. I might be overwhelmed by sensation with a million little jellyfish gluing to my skin. I might drown, or drown others, out of fear.

Somehow I managed to request the FBI's help. "I was raped once, a long time ago," I tell one of my FBI contacts. I'll call her Mary Jane. I chose a woman to divulge my secret to. A woman who carries a concealed gun on her person. She once mentioned the gun in passing, and for some reason the image of a gun stayed in my mind. Mary Jane is a tough, no-nonsense sort of person, the sort of person you would want on your team if you were involved in a dangerous mission, not the sort who would look you in the eye and offer empty platitudes, which only make the victim feel worse.

"The police have reopened the case," I tell her. "It looks like the perpetrator might have been a serial rapist. I was hoping that the police could bounce some ideas off someone in your office."

Mary Jane arranges for Lt. Macone to speak to an agent whose expertise is in violent crime. A serial rapist like this almost invariably kills someone eventually, the agent tells Lt. Macone. Lt. Macone shares this information with me by phone.

I decide I will feel about that later.

I am glad that Lt. Macone is working on solving this crime, but there are long periods when I need to be in the present, doing my "real" work. My life is busy. Almost immediately after we had a child, my marriage fell apart; or perhaps more accurately, the fissures that had been there from the very beginning became more obvious. With the birth of our child, I no longer felt able to work all the time or to comply with the "rational" way my husband chose to live. I discovered that I wanted to be outrageously inefficient—to chat about silly things with the neighbors, to bake cakes (despite the proximity of excellent bakeries), to read poems (despite their irrelevance to my work). I wanted to play the piano, even though it's far too late for me to become a professional musician.

I have been a single mother now for nearly two years. As my son gets older, I take more and more pleasure from raising him. But I find it harder, not easier, to raise him alone; and much of my psychic energy goes toward my son. My ex-husband consoles himself with the thought that I have lost interest in men, that I am focused exclusively on raising our child. My days are consumed with teaching, writing letters of recommendation, correcting page proofs, baking muffins.

Then I get an e-mail from Lt. Macone with the heading "almost done." I open it instantly, but don't find the time to answer. The subject, as usual, makes me sleepy and dull. Paul tells me he is 99 percent sure he has the crime figured out. "Please don't feel you have to pursue the details of the investigation unless you want to," he says.

I wonder whether I ought to tell Paul that I want to pursue the details with him as soon as I find the time. But somehow, the message I intend to send him doesn't get sent.

And then I got an e-mail with the heading "Done."

I am caught between two magnetic forces. Curiosity, on the one hand, and a sleepy feeling on the other. Paul has collected enough information that my curiosity wins out over my fear. The fear I don't feel. I don't tell Sara what I'm up to, or anyway, not in detail. Telling her would make it real.

I am not afraid. Nonetheless, I cannot drive out to the police station. I know that I will be too sleepy to focus, that I will get lost.

• • •

But there is a way I might be able to get out there. I have a new boyfriend. He is from Concord, too, and he knows Paul Macone well. I've known Chet for nearly thirty-five years, from around the time I was raped, but in an entirely different context. He was a family friend, ten years older than I, not someone I would ever even think about, you know, in that way. I thought of him, I would later confess to him, as a piece of furniture in the background of our family. An innocent observer, unlikely to be able to comprehend the complex dramas unfolding before him in our living room, and unlikely ever to have any interest in me. There was so much secret pain and fear in our family at that time, which we tried, all of us, to keep hidden from ourselves and especially the outside world. Every time I discover that people outside our family recognized some of what was going on, I am surprised. I thought the adults at least were able to "pass."

I had seen him five years earlier, on the shuttle back from Washington, where we had both been traveling for work. He is on the board of Oxfam and had been attending a meeting. I had been teaching at CIA University.

Seeing this person I've known since childhood, I am aware of how surprising my life path has been. This always happens to me when I see people I knew in childhood: I feel like a fraud. How could I have become this person that people consider to be an expert? Chet was a liberal congressman from our very liberal town. He is now doing what a liberal should do—serving on the board of a humanitarian organization—while I have been lecturing to spies.

Our plane was delayed. It was hot on the tarmac. We took off our jackets. I took off my shoes. This seemed permissible; Chet is a family friend. He asked about my family. I was surprised to discover that he saw through our pretenses. He'd noticed more than you would expect of an armchair, even one that had been in the family for many years.

We were sitting in the first row of coach. The wall separating coach from first class was covered with a new-looking rug. I wanted to sink my bare feet into the rug, but I hesitated. I was afraid to put my feet out there for us both to see, but I wanted to. I thought to myself that my feet must be sweaty. It would be wrong to put my sweaty feet on that rug. Wrong. Next I worried that the rug was dirty from other passengers' sweat, that the dirt would contaminate my feet. Dirty.

But I wanted very, very much to sink my feet into that rug. What you might call desire, in this case, wins out over what seems like shame and could be fear, but only in terms of my actions, the movement of my feet. I cannot resist checking in on shame and fear, the way your tongue compulsively flicks across your teeth when you've just been to the dentist. Are these really my teeth? Yes, they are. Am I still ashamed and worried? Yes, I am.

I can still call back the feeling of luxurious repose when I allowed myself to sink back in my seat, my feet at last where I want them to be, at rest.

Chet drove me home when the plane landed. But we were both married, and anyway, family friends. Neither of us expected our marriages to end. In my case, the more painful I found it to be married to my economist husband, the more determined I was to hold the marriage together. My husband was smart and funny and reliable, and our lives worked. It was true that he didn't seem to like me very much, and he certainly did not approve of me, but I had an idea that his disapproval would make me better and stronger. Perhaps surprisingly, when we finally had a child, I decided to leave that easy life in the belief that my child and I would be happier alone. Two years later, a mutual friend will introduce me to Chet, not realizing that we already knew each other from long ago.

Chet and I have a lot of fun together. He persuades me to do things I haven't done in years. We take Evan on trips for the weekend, just for fun. We take long walks. We go cross-country skiing. We attend concerts together. We are utterly, deliciously, irrationally inefficient.

I put up all kinds of roadblocks. I tell him about all my demons. From the very beginning, I let my craziness show, thinking that maybe it will drive him away. It doesn't. He's heard about me from all sides in any case, having known my family for years, from the time he was in high school. Eventually, despite my uncertainties, we become intimate friends.

It was Chet's idea to ask the Concord police for the entire file related to my rape, and he is encouraging me with this project. He has an idea that I won't be able to be intimate with a man unless I confront these demons, all the demons related to my relationships with men.

Still, can I really tell him that I cannot drive out to the police station, even though he knows that I've driven out to Concord hundreds of times? Remarkably, he takes my driving problem in stride. He does not balk. He says he will drive me, even though he has to leave work in the middle of the day.

He comes to my small apartment, and we have lunch.

No, I am not afraid to learn about my rapist, I tell Chet. Of course not. It's just that I get lost. I tell Chet that I need a chauffeur, not a therapist.

I take some care in preparing lunch for the two of us. I serve a salad of frisée, baby beets, and hazelnuts, and smoked salmon dressed with chervil.

He breaks off a piece of fatty fish with his fingers. I watch as he puts the fish in his mouth. He breaks off another piece, seemingly unaware that I am observing him. His hand looks meaty and raw to me now, with penislike fingers.

He stands up and walks toward the sink. I watch him touch the faucet with the same meaty hand. He fills the kettle. Then he takes the same hand, unwashed, and turns the burner on for tea. I see, in my mind's eye, the smear of fish oil he has left behind on the faucet. I see, in my mind's eye, a slick track of penis prints, glistening on the stove. I imagine the smell. The smell of fish oil, so much like the smell of semen.

"Wash your filthy hands," I tell my new boyfriend.

It is obvious to me now that we cannot continue this relationship. I tell him that, too.

I am not obsessively clean, as anyone who knows me would readily confirm. But this person is too crude, too disrespectful; too unaware of the tacky slick of semen he has left behind.

"You are not in control of yourself. You grab things, and you emit too much," I say. Why did I ever imagine I could be in relationship with this man? With any man? With any person?

I am vaguely aware I might be saying things that would seem odd to an outsider, things that I might later regret. But in this moment, it doesn't matter. This is an emergency. All that matters now is getting this penis-fingered person out of my apartment and out of my life.

He mumbles something about my arrogance, about the possibility that I am using him.

Now I am truly incensed. "I will not be using you," I tell him primly, haughtily. I am vaguely aware of a familiar, moralistic lilt to my voice. In my mind's eye, I strap on those high-heeled shoes, the kind with the square heels and pointy toes, the kind worn by the Wicked Witch of the West. I urge him to leave at once.

In this moment I feel myself to be dignified, self-contained.

But a part of me fears the worst, that I look and sound like a dominatrix librarian.

He doesn't leave. Instead, he walks toward me and wraps his arms around me. Loosely. He just stands there for some time, his body next to mine. There is a solidness to him. He tells me that he loves me. Right, I scoff to myself. I am not going to fall for that. I don't believe him for a second, but for once I hold my tongue. What could he possibly know about love, this undisciplined person whom life has barely tested? Never raped. Beloved mother still alive. But the energy in his body calms me down. After a few minutes, the electrical storm has passed. I am back to myself.

Chet tells me it is time to go. I hear the sound of water splashing as he washes his hands in the bathroom sink. I note the scent of lavender soap. After that I feel able to sit near him in the car. I cannot talk, exhausted by my outburst. I grow sleepier as we get closer to our destination.

When we arrive at the police station, I manage, finally, to pull a patina of sanity over my features. I am wearing a brown wool skirt, black tights, a white shirt. I think to myself, I look normal. But when we get inside, I am dizzy. Have I been here before?

Paul pulls out files, telling us what he found. A serial rapist. Many victims. Paul has found forty-four of them so far. There may be more. The victim next in line after my sister and me, he tells us, killed herself.

He gives me a large file. He warns me that it contains a picture of the perp. A picture that had been in the police files all along, in connection with the earlier rape. I simply cannot force my brain to recollect my rapist's face. My sister managed to describe it for the police immediately afterward, but I could not, even then.

The rapist, he tells us, is dead.

Dead.

I take this in.

This could adversely affect my research. Yes, I know, I ought to be relieved. But actually, I'm disappointed. This feeling of disappointment seems so crazy, so not normal, so embarrassing, that I press it out of my mind. I will feel about this later.

But now, typing up this chapter, I am ready to admit the embarrassing truth: I had wanted to kill my rapist. Not with a gun, but with the electricity in my eyes.

I will keep investigating him, of course. I need to put him in a coffin, and to do that, I need to understand him.

Several months later Paul writes again. He has an enormous box of documents from the rapist's prison file. Once again, Chet agrees to serve as my chauffeur.

In the box are several letters my rapist wrote, some written by hand, others on a typewriter. Some days the rapist was articulate. Other days he seemed to have difficulty writing. Paul wonders out loud, Could this man have had multiple personality disorder? Some observers described him as well-behaved, quiet, orderly; others as shouting, aggressive, a "pig." This trip to Concord is a different experience, however, now that we know that the rapist is dead.

Although I had trouble getting myself to come here, now that we are here and there is work to be done, I go into a familiar altered state, but different from the one where I become a moralistic dominatrix librarian. I am not afraid. I am not angry. I am interested, a spy.

It is as if this weren't my own life, as if this weren't information about my perpetrator. Even now, writing this down, I don't recognize my voice. I feel as if I were in another room from the true "I." I feel sad, locked out, bereft. Perhaps it is because I haven't yet cried about what I'm discovering.

I touch the papers my rapist touched, a bizarre act of intimacy. He is dead. I thought he planned to kill me. I see, in my mind's eye, a map with two exits. One way leads to the person I would have been had I not been raped. The other way leads to the person I became after that hour, the person I am now, a person capable of reading her own rapist's file without feeling emotion.

I take a sheaf of paper from the box. At one point, I lick my finger, trying to gain some kind of traction, determined to get through this quickly, eager to get to the heart of the matter. Suddenly I realize what I have done. A childish thought comes into my mind. Could molecules from my rapist's skin be clinging to these papers? Have I ingested a piece of my rapist? I am nauseated. But quickly, I brush the thought away. I sweep my mind clear. I become efficient and smart when I'm in this no-feeling state. The intellectual part of my mind—the accountant—does well, although I tire quickly.

I note that Paul and Chet are reading slowly. They want to talk about what they see like those annoying people who insist on reading parts of the *Sunday Times* aloud to you while you are engrossed in your own article. Somehow, despite the interruptions, I zero in on something surprising. It is a request by a woman to be informed regarding the release of the rapist. She had not filed a complaint against him, but she must have known him, and known that he was dangerous. Paul expresses amazement that I manage to find this interesting document so fast. You would make a great detective, he tells me. I will not be falsely modest: I know this to be true.

Our perpetrator was a madman, it seems. Much of the time he was passive, quiet, even polite, according to the reports; sometimes extremely odd. He was observed standing, stark naked, on his toilet. He was observed eating paper. He committed many petty crimes.

He was not popular with his fellow inmates. They torched his cell. Child rapists are despised—not just by their victims, but also by other criminals. He was put in protective custody.

I find a progress report on the inmate. My inmate. My rapist. On January 28, 1974, he was in protective custody at Walpole.

"He has spent very little time in the prison population due to the nature of his offense. The team and the resident concur that counseling is of the highest priority, in that the resident might better come to grips with the length of his sentence. At the same time, it was agreed that the resident would most profit from a comprehensive program that would enrich his skills as a plumber."

His skills as a plumber. He did indeed plumb me. The resident, it appears, was obsessed with plumbing.

The resident? A nauseating image comes to my mind, the rapist as a medical resident, performing Mengele-like experiments on the plumbing systems of little girls.

I have always found it dizzying when people refer to our innards as "plumbing." Such a vulgar expression. But we are beyond vulgar here. We are in the muck of terror.

I see now that this man didn't just reside in my mind's eye; he resided on this earth. He was apparently a fellow human.

I find, in these documents, that the perpetrator was from a small town, Milbridge, Massachusetts.

Brian X. Beat
1 Orton Street Milbridge MA
Date of Birth 2/5/47
Height 6'2" Weight 155 Blue eyes Hair brown

Mr. Beat attended school in the Worcester area and while a student in the 11th grade at North High School, Worcester, he dropped out for a year. He was unable to remember why and could offer no reason to this dropping out of school.

He returned to school the next year to complete his secondary school education at Milbridge High School. He graduated there in 1965. He has a 4F draft classification and explained that this was given because he was determined as not suitable for military service because of his admitted homosexual activity. In explaining this further, Mr. Beat indicated that his homosexuality lasted only until he received the 4F classification. He also explained his homosexuality as an early developmental problem. Since then he states that he has only had sexual relationships with women. He also tried to enlist following his 4F classification but was refused. Mr. Beat was adopted. While an adolescent he was seen by a psychiatrist in Worcester, a Dr. Roy. He was referred to this psychiatrist by an aunt who said he needed someone to talk to, someone to help him get oriented and to help him establish his own goals.

The defendant claims to have attended a parochial school, St. Louis Elementary School, in Webster, Mass., for grades one through six, Woodard Jr. High School in Milbridge, Mass., for grades 7 through 8, North High School in Worcester, Mass., for grades nine through eleven. He dropped out of school for one year and continued to complete his twelfth year at Milbridge High School, Milbridge, Mass., in 1965.

At this point, Lt. Macone is sharing this information with me because he wants to help me. Now that we know the rapist is dead, Lt. Macone no longer feels the need to find him. Brian Beat cannot commit any more rapes. Now, if I cannot resist learning more about my rapist, I will be on my own.

• • •

Lt. Macone gives me the name of one of the detectives who knew Beat in Milbridge after he was released from prison. Detective George Remas.

After hesitating for a couple of months, I call Detective Remas. He does not return my call. I call again. He doesn't respond. I ask my research assistant to leave messages every day. I guess they are busy out there. Eventually I write to the chief of police, thinking that perhaps Detective Remas is afraid to talk to me without his chief's permission. That does the trick. Finally, we have a date.

Chet and I drive out to Milbridge. It's only an hour from Boston, but you feel like you're in a different state. Iowa, maybe. Milbridge was a mill town. Later it was known for its tanneries and shoe manufacturing. The mills and tanneries are now gone, and the town feels dead. It is utterly flat. Local lore has it that Lindbergh once landed a plane here, and that they turned the airstrip into a town. I have no idea whether this story is true, but the idea that this flat town was once a landing strip captures the mood well.

The police station is right on Main Street. It looks like a small suburban tract home. Perhaps they don't have much crime here, I think to myself.

Now that we're here, it turns out that Detective Remas is friendly. He seems almost relieved to have time off from duty. He did not know Brian Beat before he went to prison, he tells us. But he saw him frequently after Beat was released.

"I was always arresting Brian because he was throwing trash at cars, or he was drunk, or because he committed a crime," Remas tells us.

Beat lived in a kind of alcove right next to the railroad tracks, he tells us. He built fires there. He collected things: bags of cigarette butts, the bar codes at the bottom of Marlboro cigarette packs, news stories about accidents. The police would know where he had been because they would find his collections.

"Sometimes he would give his clippings to me. News stories about police-cruiser accidents. He would hand me a clipping and say, 'Kojak. I got this for you.' He knew my name, but he always called me Kojak," Remas says.

He wore tattered clothing. Pants on top of sweatpants and shorts over that. Gloves with holes in them. Or he'd use an old pair of socks as mittens.

"One year I gave him a coat, an old coat of mine," Remas says.

He pauses, his lips slightly pursed, as if he is not sure whether to continue. But he can't stop himself. It must be painful to have a convicted rapist of children return to your small town.

"I told him I had something for him, that I wanted him to have my coat. At first he refused it. Finally he accepted it. I saw that he was welling up. That was the only time I saw him show any kind of emotion."

He seems to feel the need to defend this act of generosity.

"You get sick and tired of chasing the same person day in and day out," he says.

Suddenly he seems to recall the reason that I'm here.

"I knew from his record that he was a sex offender, but I never heard of him committing that kind of crime here."

He is speaking haltingly now.

"Do sex offenders change?" he asks.

Am I supposed to voice an opinion?

After a slight pause, he answers his own question.

"I don't believe so," he says.

I am deeply relieved that he found a way to answer himself.

But Brian Beat, he tells us, was never put on the list of registered sex offenders. I wonder why, but am afraid to ask.

The subject of sex offenders reminds Detective Remas of another topic.

"There was a certain group of kids that would pick on him. They would antagonize the poor guy. But there was another group that got on well with him. They had an arrangement: they would give him money to buy a case of beer. Then they would offer him a few beers as payment. He'd walk out of the package store and you'd see kids farther down the road."

Were any of those kids girls? I wonder. I don't ask, afraid that if I do, he may stop talking. It is probably not a good thing for a police department to reveal that officers had witnessed a convicted sex offender buying beer for local children.

"Brian never admitted to anything." What crime does Detective Remas wish he admitted to, I wonder?

"I never heard Brian say he was sorry. I never saw him sober," he adds. But then he adds, "He was a haunt. A pest to the community."

He would walk up and down Main Street, Detective Remas says. Sometimes Remas would see him walking up and down the railroad tracks at night. He often built himself a fire under the bridge, where it was dry. He would jump in the river to wash his clothes, and dry his clothes by the fire. Sometimes he would set his clothes on fire, Detective Remas explains. But he never bathed himself. He reeked, Remas says.

"When it was cold, he would sleep in the bathroom in the gas station. Stevie's Global. Right down the street. Stevie let him sleep in there. He would use the sink to wash up. Did him no good though. The crud on his ankles," Remas tells us.

His arm passes across his mouth in disgust at the thought of that crud.

"Brian and Stevie were friends," he says, apparently not surprised that a convicted rapist of children would have friends in town.

"If it got to be too cold or he wanted something from us, he would stand at the corner of Main Street and Harwich, throw shoes or clothing or trash at passing motorists. Always at that corner. When he came in here, the sergeant would make us take his shoes off outside because of the odor. But his feet stank, too. He would urinate and defecate and ejaculate all over the walls. If we didn't have Plexiglas over his cell, I'm sure he would have been throwing excrement at us."

And with a sigh, he summarizes his assessment. "Brian was a pig," he concludes.

"He could be very calm and talk to you about history, next minute he would start screaming at you. 'I'm going to call Johnny Cochran,' he would say. I tell him, Brian, here is the phone, go right ahead and call him. Then he would go right back to asking what time supper is being served. 'I want a bologna sandwich with onion.' He always wanted a bologna sandwich with onion."

Once again, Detective Remas seems to recall why I'm here.

"Somebody with a background like that. You don't rape children or molest someone just like that. Something must have happened to him. If I get a confession in a sex-crimes case, I always ask, has anything like this happened to you in the past? The answer is always yes. We don't just wake up one day and molest children. Something happened to this guy. It is a learned behavior. The idea to rape children didn't just pop up in his mind."

Then he adds, "We had a priest at Saint Roch's . . . ," but he doesn't finish the thought.

He starts again. "We had other men in town that had been molested. They were in the south end of town."

Was Detective Remas telling me that he thought Brian had been molested by a priest? If so, I don't think he intended to.

"When we arrest someone, we have to list properties that they come in with. Brian always had a lot of stuff in his pockets. He'd collect cigarette butts. In his pockets. Empty matchboxes. He kept plastic ware, pennies. It would take you a while to empty his pockets. Always at the bottom, small stones. Wrapped them in tissue paper. Dime-size pebbles wrapped up in the tissue paper. A dozen of them. Tissue paper with little pebbles. I'd throw them in the clear plastic inventory bag. I would ask him, What are you doing with the rocks? He would just say, 'Those are my property.' Always in tissue paper. It wasn't significant."

Detective Remas is content to talk, stream of consciousness. But now Chet has a question.

"Were the pebbles ever in his back pocket?" he asks.

"No, never in the back pockets. In the back he would have plastic bags. A filthy comb. Maybe an extra pair of socks or the gloves without the fingers. The pebbles were always in his front pocket."

"Can you tell me more about the priest?" I ask. "The one from Saint Roch's?"

"Father Sissen," he says. "Last I heard he is blind and living down on the Cape."

"Was there ever a formal complaint against him?" I ask.

He doesn't know. The church had relocated him. And then there was another priest after Father Sissen. "Father Smith," he says. "Same nature."

How could the church send two pedophile priests in a row to this tiny town?

"They were very quiet about it, but I went to school with one of the guys who was a victim. I know two of his victims. Two guys. They still live in town. One of them is very vocal. Then Pat Hanley, the new priest, same problem. Three priests at one church. All pedophiles."

Is this really possible?

"But the kids loved Father Sissen. He would hold dances. He

had a summer home on the Cape, and the kids would go with him. He treated the kids good. He would mingle with the kids. He went to every basketball game."

"Did he use rosary beads?" Chet asks. Detective Remas doesn't seem to notice the question. His thoughts have returned to Brian Beat.

"Even when Brian was drinking, he would recite poetry. Riding back in the cruiser, he would go on and on. I would turn up the radio. Brian, enough's enough. There was one poem about the Garden of Eden . . . Brian always carried paperback books with him. Always had books. When he came in here he wanted the papers. He wanted the *Globe* and the *Herald*. He would sit and read all weekend long. When he was done, he would throw the papers in the toilet. He would cover the vent or the light in the cell."

"What books did he read?" I ask, slightly repulsed by my own curiosity, an impulse that Detective Remas, apparently, does not share.

"I couldn't tell you what they were, the books he read. I was looking for razor blades hidden in his books, not at the titles.

"I remember the day we got called to the house when he had hung himself. He had taped some poetry on the closet door. He had a thing about poetry."

Detective Remas is visibly upset. Something about the death of Brian Beat troubles him deeply. Is this the reason he didn't return my calls?

"His mother. She had a routine. He knew that she was going to his father's grave that day. He knew her routine. . . . He usually made a pot of coffee for his mother, left it there for her to drink when she woke up. But that morning there was no coffee. This cold son of a bitch hung himself right where he knew she would have to run into his dead body as she went to her car. She told me, 'I was going to the cemetery to lay flowers out. He knew that I was going that morning.' I'm the one who had to cut him down."

So Remas had to cut him down. That is why this topic is upsetting to him. Or one of the reasons.

Detective Remas hands me some photographs that he took at Brian's mother's house on the day Brian hung himself. There were poems, and a picture of the game called Hangman.

The first poem appears to have been typed on an old typewriter. The letters are uneven.

"Before,
Nobody had
Wanted me to
Take my life in
My own hands.
I felt I finally
Had to find my
Own identity
And what I
Really value."

A second poem looks as if it was torn from a book:

Don't be Afraid to Fail
You've failed many times, although
You don't remember.
You fell down the first time you
Tried to walk.
You almost drowned the first time
You tried to swim.
Did you hit the ball the first time
You swung the bat?
Heavy hitters, the ones who hit

The most home runs, also strike out a lot.
R. H. Macy failed seven times
Before his store in New York caught on.
Novelist John Creasey got 753 rejection slips
Before he published 564 books.
Babe Ruth hit 714 home runs, but
He also struck out 1830 times.
The message is, don't worry about
Failure.
Worry instead about the chances
You miss when you don't even try.

I find an almost identical poem on a Web site maintained by a person aiming to inspire writers, and on another Web site that offers prayers.

Can I stand the feeling of feeling sorry for my perpetrator?

Later, Brian Beat became a suspect in the abduction and murder of a girl who lived near his parents.

I had never heard of this case; it's the sort of story I would avoid in the papers. Too awful. I look her up and find a picture of her online.

I see a picture of a blond girl. The words *aggressively innocent* come to my mind. At first I feel her look is challenging, as if she is daring us to notice her smooth skin, flushed cheeks, half-open rosebud mouth. I sense that she may have just come into awareness of the power she might have over men. I can imagine that mouth forced open, and I hope that unlike me, she had the presence of mind to bite very hard. The teeth look strong. She was an athlete. The lips are red, suffused with energy. I notice now that the mood in her face is divided. If we look only at the mouth and chin, we could be looking at a grown woman. But her eyes have a calm innocence. A provocative innocence, but innocent nonetheless. There are two glints of light in each eye. The light above her pupil is the light of her soul, I think to myself, absurdly. The one below suggests a tear, as if she were crying out to me to help her, to help little girls trapped in women's bodies. Her skin is silken, suffused with a red flush, the look of a fairy-tale princess, Snow White perhaps. Her pupils are strangely wide, given the light in her eye.

Is this what I looked like back then, aggressively innocent?

chapter four

The Making of a Spy

As I said, I have always been a spy. I started with my own family. Our family life was defined by mystery, and I was determined to pierce what lay beneath the mystifying miasma. I have learned, again and again, that behavior that seems cruel on its surface might actually be motivated by forces other than malevolence or selfishness. Behavior that seems kind might actually be cruel. I had an urgent need to understand people's hidden motivations, as if it were a matter of safety. As if the viability of my family were at stake.

Discovering that my father was waiting behind the door when gun-wielding Nazis were threatening his mother made my father's decisions easier to understand. When I asked my father whether he thought that it was possible that his mother was raped behind that closed door, he said, "She had washer-woman knees. No one could possibly think of her as a sexual object. Besides," he explained, "she would have told my sisters, and they would have told me." I am not so sure. Maybe someone needed to ask her. Someone needed to want to know, to be able to bear the answer.

I discovered early on, long before talking to terrorists, that some secrets are relatively easy to pierce. Few people seem to realize that often, all you have to do is ask. You need to pose a question and then sit silently, your whole being focused on the speaker and her story. If you are genuinely curious about people's inner lives, you will learn amazing things, stories better than Tolstoy's. You must watch for movements on the periphery of your vision, thoughts on the periphery of your mind. If you watch closely and listen attentively, secrets will be divulged—whether purposefully or inadvertently. You have to be deeply curious for this to work, of course. I am blessed, or cursed, with curiosity, and as a result I have learned a great deal.

For example, I learned, simply by asking my grandmother, that my grandfather was a roué. This knowledge came about as a result of a picture I found in my grandfather's desk. I don't know why my attention was drawn by the picture. I had no preconceptions. "Who is this?" I asked my grandmother, pointing to a lady with very pale skin and dark hair. I was naive, open to whatever story my grandmother might have to impart. "Oh, that. She was your grandfather's girlfriend during the war," my grandmother replied, with little emotion, as if I had asked her about an unfamiliar hat. I was puzzled. Didn't she love my grandfather? Why wasn't she bothered by my grandfather's affair, or by his retaining a photograph of his lover? My grandfather was a "dirty old man," the words we used, when I was a girl, for an old man who seemed inordinately and inappropriately fascinated by sex. He told dirty jokes. He talked about women's bodies. He said "Ooh lah lah" in response to clothing he found attractive, even clothing worn by his own granddaughter. But an affair? This was hard to believe. When I was a girl, I thought adultery was something you read about in books, not something that occurred in your own family.

"He thought he was madly in love with her," my grandmother continued. Do I remember this correctly—this "thought he was in love?" He was so in love with her, my grandmother explained, that he considered remaining in England after the war, leaving my grandmother and my then twelve-year-old mother behind to fend for themselves.

I think now of my grandmother's face, her eyes so soft and brown and full of love that you could rest your soul there, and almost feel safe. My grandmother was astonishingly beautiful. She believed in taking care of herself. Going to her house was like going to a spa. She would urge us to use her luxurious creams, her Vitabath, her jewelry, her makeup—always the finest that could be had. Her mink coat always smelled of Joy perfume. She had a large collection of Ferragamo shoes. How could this have happened to her? How could my grandfather, who clearly loved her, have betrayed her in this way?

Still, my grandmother spoke calmly. Wasn't she worried about raising my mother, her beloved Shola, on her own? I kept this information in the files of my mind, unbelievable as it was, to explore further at a later date. I knew that my grandfather adored my dead mother. There were pictures of our mother all over our grandparents' house, unlike my father's house. Many of these photographs were taken by my grandfather. He was a talented photographer. He even won a contest, my grandmother said. Some of his photographs were published in the newspaper. The house had become almost a shrine to our dead mother.

The unburdening of this secret about my grandfather's love affair led, inevitably, to the revelation of others. There was my grandfather's nurse, Anita, who sat in the entryway to greet my grandfather's patients when they arrived at the doctor's office he maintained in a separate section of our grandparents' home. She performed many duties for my grandfather: receptionist, nurse, bookkeeper. Like my grandfather's wartime lover, she had jet black hair and white skin, and looked remarkably like a painting that hung above the stairway leading to our bedrooms.

My grandfather, as we knew him, was not a loud man. He was sometimes ornery but usually quite reserved, even taciturn. Every day, at precisely noon, he would wander to the back side of the house, where family activities took place, and sit down to a formal lunch with whoever happened to be home that day—my grandmother, the occasional relative or friend, and, during the period when my mother was ill and dying and for the year after her death, my sister and me. But occasionally, after the patients had left the office, just before lunch, or at the end of my grandfather's workday, just before dinner, we would hear shouts from behind my grandfather's office doors. My grandfather and Anita fought in a way that my grandparents never did. "He fights with Anita so he doesn't have to fight with me," my grandmother told us. Later she added that my grandfather and Anita were lovers.

Anita, like my grandparents' loyal maid, Jean, was a seemingly permanent part of my grandparents' household. She served as my grandfather's nurse from the earliest period I can recall to the end of his medical career, by which time I was in my late twenties. But my grandmother told me that there had been predecessors to Anita, some of whom fulfilled a similar role.

I loved my grandfather fiercely. He was harder to love than my grandmother, and he needed a champion. I loved him in spite of—or perhaps because of—the many aspects of his personality that were hard to bear or defend. He was a male chauvinist, or so it seemed to me. He leered at girls. He was also a racist. I wanted to show him that I was at least as smart as he was, even though I was a girl. And I found ways to torture him for his racism, as did my mother, I would later find out.

Here is what I loved about my grandfather. He was bright, known for his uncanny ability to diagnose difficult cases. I was proud of this. He was serious and suspicious and terribly judgmental (though not, apparently, of himself). You had to earn his love and respect. This was very much in contrast to my grandmother, who seemed to love everyone, and whose love, therefore, seemed cheap. My grandmother once informed me that I, the more serious, dark-haired and dark-eyed older grandchild, "belonged" to my grandfather; while my sister, the more cheerful, light-haired and light-eyed younger grandchild, "belonged" to my grandmother. I wasn't sure what to make of her statement.

When I was twelve or thirteen, my grandfather suddenly decided that my nipples were dry. I am not sure how he managed to see my nipples, or how he reached this surprising conclusion. He sent me to a doctor, a close friend of his. The doctor prescribed a regime that involved washing my nipples with alcohol, followed by the administration of Vaseline. This supposedly curative regime was imparted not directly to me but via my grandfather, who spoke with the other doctor by telephone. There was something unnerving about the look on my grandfather's face when he spoke to his friend—a kind of conspiratorial look. At the time, I wondered if this doctor had been involved in diagnosing my mother's cancer. Now I think it might have had something to do with my grandfather's sense of ownership over my breasts. I cannot remember what happened after this breast-healing plan was imparted to me. But in any case, the system of which grandparent owned which grandchild eventually broke down, I'm not sure why, and my grandmother and I became very close.

After my grandfather died, I questioned my grandmother further about my grandfather's lovers. It seems there were many of them. "Weren't you jealous?" I asked her. "No," she said. She knew that my grandfather loved her, she said. And that was that.

I found her equanimity in the face of this insult impossible to fathom.

Something made me ask her whether she had ever taken a lover. "Yes," she said breezily, as if I had asked her whether she had ever tasted one of Zaro's glazed doughnuts. Zaro was one of my grandmother's "friends," their relationship solidified by his willingness to donate money to my grandmother's favorite liberal causes. My grandmother's whole life was politics. She served, for a time, as the chairperson of the Democratic Party in Westchester County. My grandmother was a do-gooder, but never sanctimonious, and people seemed happy to be charmed into helping her. She loved to serve Zaro's bagels and breads at political functions and fund-raisers.

"During the war," she told me, "a man fell in love with me. He wanted to marry me. He wanted to adopt your mother," she said.

I was stunned. Too stunned to respond, unusual for me.

"He painted that picture over there," she said, pointing to the familiar painting hanging over the dining room table. That picture, which had been there as long as I could remember, was now alive with new possibilities, a totem from an alternate life, in which my mother would have been raised by a man who was not my grandfather. It is a picture of a lone woman, standing next to a chair by a yellow-gray sea, staring at the horizon. The picture hangs in my sister's dining room now.

"That was a portrait of me, waiting for your grandfather to return," she told me. I had never thought to ask who was portrayed in that painting, which I assumed was a copy of a forgotten masterwork, like so many of the paintings in our grandparents' house.

Why would this man, this painter, want to adopt my mother?

My mother already had a father. Her father was still very much alive at the time, even if far away, treating soldiers who were wounded in the war. Did he intuit that my mother needed protection?

I pressed for more details. He was a psychiatrist, it turned out, not a professional painter. My grandmother met him at one of the community meetings she regularly attended.

"Did you love him?" I asked.

"Not really," she said. He satisfied a need at the time, apparently, rather like Zaro, the owner of the bakery.

"Did you feel guilty?" I asked my grandmother.

"No," she said, apparently surprised by my question, as if she had never heard that betraying one's spouse is generally considered to be a moral failing. For my grandmother, a parking ticket always meant a call to the city manager, who would "take care" of it for her; a long line at the post office meant sweet-talking her way to the front. This was the biggest puzzle of all—my grandmother's apparent freedom from both anger and guilt. She seemed—is it possible?—to lack a conscience. It was as if she lived in a bubble where the normal rules about fairness and integrity did not apply—either to herself or to those she loved. In my grandmother's world, love was all. This was her way of coping with what life offered her. Although she answered all the questions I knew how to ask, I was never satisfied that I understood her, plagued as I am by shame and guilt, even for crimes I have only contemplated.

But somehow I understood that this was the end of the conversation, that my grandmother had revealed all that she ever would on this topic.

Some secrets, however, required greater effort to pierce than simply asking questions.

There was the mystery of my mother's life as well as her death, which was difficult to unravel by posing questions because I was afraid to bring the topic up. Fortunately, sometimes people offered information, apparently intuiting that I might like to know. My father was not one of them.

I was told by my grandparents, who idealized their dead child and who could not be relied on as objective witnesses, that my mother was remarkably brave. Other relatives spoke about her liveliness, quick-wittedness, and rebelliousness. She was politically aware from a young age, apparently following in my grandmother's footsteps. One of my mother's acts of rebellion was to attend her 1947 high school prom with a black man as her date. And now I wonder: Who was that man? Was she in love with him? Where is he now?

My grandmother, who was involved in civil rights, insisted that my grandfather treat black people, whom he otherwise might not have included in his practice. She also talked him into treating poor people without charge. The poor people, in those days, were a mix of immigrants and blacks. Although my grandfather was riddled by prejudices, he would often soften when confronted by another person's physical pain. But I could not stand his prejudices. Like my mother, but before I knew her history, I made a point of bringing a black boyfriend by to meet my grandfather.

My mother wanted to become a labor organizer. She had finished high school at age sixteen. She applied to the Cornell School of Labor Management but was rejected on account of her youth. Her second choice was Duke, so she started college there. At the time girls were forbidden to walk on a certain path at night, out of fear that the "niggers" would get them. But my mother organized a candlelit march on that very path. Soon after that, she was expelled. My grandmother told us that she then wrote to Cornell to explain what happened, and Cornell accepted her. My parents were both students at Cornell when they met.

My grandfather's character was quite different from his

daughter's. He was fretful and fearful. He worried constantly about my mother's various childhood illnesses, as he would, later, about my sister's and mine. Because his office was in the house, it was easy for him to treat family members at any time of day. He was fascinated by the power of antibiotics. My sister and I have gray teeth as a result of his frequent medical interventions. He was an excellent diagnostician, or so it was said. But he was so clumsy that his lancing of an infected toe or a blister could result in a deep wound. There is an indentation on the top of my right big toe. Probably I was born with it. But for some reason, I have an idea that my grandfather accidentally gouged me there; that he had meant to remove a wart or perhaps even cut my nails, but instead removed part of my toe. To this day, when I feel alarmed for any reason, I hide my feet under my body.

Despite these painful memories, or perhaps fantasies, I found my grandfather's office irresistibly intriguing. Throughout my childhood and teenage years I would regularly retreat to his offices when my grandfather wasn't working, with a novel or homework in my hand. It was peaceful and quiet in there, beyond the double doors that separated the office from the rest of the house.

My grandfather's office consisted of four separate spaces—the waiting room, the examining room, the X-ray room, and the consulting room. There were doors everywhere. It was a world apart, but connected to the house through two sets of double doors.

The waiting room had wall-to-wall carpeting in a cheerful royal blue, and an airy, uncluttered feeling. It felt luxurious to me, but also alien. There was a white display case with issues of *Life* and *Good Housekeeping*, and a big white book filled with pictures of birds. I didn't stop to spend a lot of time in there.

I usually went directly from the house into the consulting room. I liked to spend time in that room because you could sit at my grandfather's desk, and there was interesting information in there. The room smelled of leather, wood, and ink. There was very little natural light, but there was a glow from the oriental lamps. It had the feel of an internal room, with doors leading in three directions—to the waiting room, the examining room, and the rest of the house. There was mocha-colored rice paper on some of the walls, and paneling on the others, giving the room a cozy feeling. There was a fireplace on the wall to the right of the desk. My grandfather's massive, carved oak desk stood in the center of the room, with shiny, pounded nickel pulls.

I knew where everything was in my grandfather's drawers—the checkbooks in the top right drawer, the envelopes and stamps in the second, paper clips and tape in the third, and the bank statements in the bottom. The drawers on the left contained financial records—taxes and statements from Smith and Barney. I'm surprised how many details I remember because I am generally not visual. I remember far more about what was contained in my grandfather's desk than I know about the contents of my own desk now.

On top of the desk—which was always pristine—was a leather blotting pad. My grandfather's two fountain pens stood primly in a green marble stand. Grandpa's chair was covered in brown leather that squeaked when you moved your body, and emitted a comforting, horselike smell.

Grandpa kept six-by-eight-inch file cards in ornate wooden boxes on the bookshelf built into the wall behind his desk. That was where he kept his private notes on his patients. I liked to sit at my grandfather's desk, remove the boxes from the shelf, and place them on the blotting pad. I understood, as a spy in training, that I had to put everything back exactly as it was, so I pulled the cards out carefully. His penmanship was flowing, romantic, almost artistic, quite out of keeping with his character, or at least the character I knew. I would study his notes, curious about the diseases his patients suffered from. Often I looked up the patients' illnesses and their medications in the reference manuals, which were also kept on the bookshelf behind the desk. The books were filled with glossy pictures of diseased organs, swollen limbs, and colorful pills. One of my favorite pastimes was to compare the side effects of medicines listed in the *Physicians' Desk Reference* with those listed in the *Merck Manual*.

The full files for my grandfather's patients were kept in the front office, where his long-suffering nurse, Anita, greeted patients, but I was less interested in those—I wanted to see my grandfather's personal notes. I have no idea whether my grandparents knew what I was actually doing when I retreated to the quiet of my grandfather's office to read. Grandpa must have suspected that I wasn't only reading Jane Austen. After all, I knew where to find stamps and tape and even his checkbook if he needed these things on weekends. Did he know that I was actually spying on him?

Fortunately, from the standpoint of patients' privacy, there was only one patient, Miss Celia, whom I knew by sight. Miss Celia was extremely old. She and her sister owned the candy shop down the street, and lived together in rooms above the store. Once I heard my grandfather telling my grandmother about her high, firm breasts, which he insisted had remained youthful into her old age because she was a virgin. There were no Health Insurance Portability and Accountability Act (HIPAA) rules back then, and Grandpa often spoke about his patients at mealtime.

The consulting room opened into the examining room. This room was filled with a cold, clean light. The floor was covered in pale gray linoleum, which Jean, the housekeeper, kept so polished you could see your reflection. When you entered the room through the patients' door, you faced a wall of glass cabinets. I knew precisely what lay in those cabinets, though I suppose the patients, fortunately for them, did not.

When my grandfather first started practicing in the 1920s, general physicians did nearly everything themselves. They delivered babies, set limbs, took their own X-rays, examined their own slides to check for various infections. There was an extraordinary array of sharp metal implements in those glass cabinets, which, earlier in his career, my grandfather must have used to perform minor surgery. There were hypodermic needles, medications in glass vials, an ancient microscope, glass slides in glassine envelopes. A stainless-steel scale for weighing babies stood on a shelf in the corner next to an autoclave. The examining table stood coldly in the center of the room. The stirrups looked to my child's eye like implements for torture.

The most interesting and frightening item in the room was the ornately decorated X-ray machine, which stood in the corner. I have no idea how it worked, but there was a kind of central operating system on the floor of the examining room—a large metal box with dials on it, the size of a small trunk set on its end. I imagined my grandfather standing behind it and manipulating the controls, like the Wizard of Oz. After my grandfather died, a medical museum asked us to donate it, and we did.

A side door opened to the X-ray room. That was where the patient would be instructed to lie down on a narrow wooden bed. The thin mattress was covered in a white sheet. I remember the smell of the sheet and its rough, starchy feel. Jean must have used a lot of starch on it. On the mantel was a clock or a timer.

I did not like that room. It was always dark, even with the fluorescent light on. It smelled of chemicals, and maybe of fear.

The patient would be instructed to lie on the narrow wooden bed while my grandfather operated the controls from the next room. Across from the bed was a deep sink, where my grandfather developed the prints, and a large file cabinet filled with pictures of patients' insides. This was the one area of my grandfather's offices that I didn't study in detail. I knew there were X-rays filed, in alphabetical order, in the metal cabinet, so I must have looked inside it once. But, unlike nearly everything else my grandfather touched, I never looked through the hundreds of prints in that cabinet. I never bothered to determine whether there were X-rays of Miss Celia's chest, for example, to see whether I might discover what her supposedly firm breasts looked like on the inside. There was something dreadful in the air here, some unnamed evil that squelched my normally insatiable desire to spy on my grandfather.

It wasn't until I was in my late teens that I heard about my grandfather's love of medical technology, and especially that X-ray machine; and his conviction, in the early days, of the curative power of radiation. Grandma and I were sitting on her four-poster bed, where I often sat with her in the morning before breakfast, or before bed in the evening, up until the day she died. It was my habit to wear my grandmother's gowns when I slept at her house. She had so many of them, all pressed and folded neatly in her second drawer. You could choose among pastel yellows, pinks, or blues, with lace or without. I don't recall what Grandma was wearing, but I do recall that I was wearing one of her pale pink gowns, which felt luxuriously smooth on your skin because it had been worn into a silklike softness. I do not recall how the conversation began, but somehow we got into the topic of my grandfather's mad-scientist obsessions. If 10 milligrams of a certain medication were typically prescribed, Grandpa would want us to take twice as much—only the best for his family. Your mother had frequent colds as a child, Grandma said to me. Or so it seemed to Grandpa. He became convinced that her thymus was enlarged, and that that was the source of the problem. So he decided to do whatever he could to shrink it. He commenced a course of radiation therapy, which he administered himself.

I would later learn that the practice of irradiating children whose thymuses were suspected to be enlarged was common at that time. The problem of an enlarged thymus turned out to be a myth, but some physicians continued the practice into the 1960s, though it was most common between the years 1924 and 1946. Even croup was sometimes treated with radiation. In those days, no one really understood the dangers of radiation therapy, and typical dosages were hundreds of thousands of times higher than we would use today. It would not surprise me, given my grandfather's habitual enthusiasm for overmedicating members of his family, if he used a higher dose than was the norm.

At twenty-six, my mother contracted lymphoma. At twenty-eight, she was dead.

An intrusive thought comes into my mind as I type. I see my grandfather's revolting body, the rough black-gray hairs on his chest and arms, his skin repulsively white under the hair. He smelled foul. Even his genitals come to my mind. When I had croup as a child, he created a tent and held me inside it, held me in his repellent arms. Later, he took me in the shower with him and held me close to his fat, naked body. Why?

My grandmother imparted this information about my grandfather's irradiating my mother to me with a deep sadness. But the sadness seemed to be for my grandfather, who suffered so much, she told me, as a result of his mistake.

I took this information in, but I did not really believe it. I told myself that it was clearly a coincidence—Grandpa happened to have irradiated my mother, and she happened to have died of cancer. Even if it were true, I felt myself to be way too young to hear about it. I put the information away, to consider at a later date. It seemed imprudent to mention this story to anyone else, uncertain as I was about its veracity. So for a very long time—for more than ten years—I kept the story to myself, shunted away somewhere in a mostly inaccessible part of my mind.

It took quite a while—perhaps fifteen years—for me to ask my father about my grandmother's tale. It's not that I was afraid, exactly; it's just that it never seemed to be the right time. But then, suddenly, it was. My father and I were alone in the car, returning from an uncle's funeral. It was dark, so I didn't have to look at my father, and he wasn't able to see me. Plus, death was in the air between us, so it seemed okay to broach the topic of my mother. I told myself I would find a way to change the subject as soon as I could. All I wanted to know was just this one thing: Did Grandpa accidentally kill my mother?

"Is it true that Grandpa's irradiating Shola is what killed her?" I asked. My sister and I often referred to our mother by her name rather than as our mother. Somehow that made our mother's life and death less real to us, enabling us to skirt around the feelings of grief and loss, feelings I knew to be exceedingly disturbing to my father.

"Yes," he said. "It is quite likely that your grandfather inadvertently killed your mother." Before I could stop him, he repeated a story I had heard before, a story that for some reason was not "off-limits" for my father, even though it concerned my mother's death. When my mother was nearly dead, my grandfather insisted on trying every possible therapy. "She was emaciated," my father said. "In agony. Grandpa was crazed with grief. He could not bear to see his only daughter die. So he insisted that she undergo this crazy therapy, that she serve as someone's guinea pig," he said bitterly. "After she died, your grandfather kept seeing her ghost," my father told me, disgustedly, not for the first time. I could tell that my father wasn't sure what to make of this phantom, or what it meant about my grandfather's sanity. I have an idea that his revulsion is a substitute for rage.

Later, I asked an uncle, a psychiatrist. "We all assumed that your grandfather's radiation treatments killed your mother," he confirmed. An older cousin said the same thing. To this day, I have trouble grasping the idea. It doesn't compute. I can only believe it some of the time. Until recently, I didn't discuss this matter with my sister. I didn't think she'd believe me. After all, I could hardly believe the story myself.

I have learned that many people who received radiation treatments for benign conditions, such as supposed enlarged thymuses, ultimately succumbed to cancer. But I'm still skeptical: correlation does not imply causation. The idea that a swollen thymus caused disease in childhood was a widespread misapprehension among medical professionals at the time. So it would make sense, given my grandfather's character, that he would want to do his best for his daughter by irradiating her with very high doses. Still, I cannot get it.

A number of things would make sense were I to allow myself to believe this story—my grandmother's apparent acceptance of life's blows, my grandfather's depression, my father's virtual disappearance after our mother's death, his loathing of my grandfather. Nonetheless, I want to hold the news at bay. It still feels like news to me even now, even though my grandmother told me these stories more than thirty years ago.

I have to wonder if I was still spying on my grandfather when I studied chemical weapons and terrorism. When I worked at a nuclear-weapons laboratory, surrounded by "mad scientists" of another sort. Later, when I was working on nuclear terrorism at the National Security Council, I was invited to visit a former Soviet nuclear site. To protect us from radiation, the Russian scientists offered us white medical coats, which of course would not protect us from anything. Some of the members of the team were a bit alarmed. But I marched right in, with only the white coat and hat to protect me. Perhaps I was taming isotopes.

chapter five

The Rapist

I know my rapist is dead. But I still need to understand him. If I can understand him, I can put him away.

I want to talk to Stevie, the owner of Global Gas, whom Detective Remas of Milbridge told me about. But every time I direct myself to dial 411 to ask for the number, I am stymied by the image of Brian Beat's dirty feet, which float, unbidden, into my mind. Why did he wash his feet in the toilet? "Did him no good. The crud on his feet." I swat at Detective Remas's words, but his words float back like unruly feathers.

I mean to go out to Global Gas. But the effort to extinguish the image of Brian Beat's filthy feet exhausts me.

Again and again I shove those feet away from my face. But they flit back. Since I cannot seem to dial 411, I ask Jack, my research assistant, to go find Global Gas. I do not tell my RA what I really want to know, which is whether it is possible to steer clear of the magnetic pull of that bathroom.

Brian Beat's dirty white feet are now dipping themselves daintily in the toilet bowl and then flitting about the body shop, refusing to stay closeted in the bathroom. I want to send a dog after them, to shred them into unrecognizable meat. Brian Beat should learn to keep his feet to himself.

My RA, Jack McGuire, returns with such compelling news that I am distracted momentarily from the image of those fluttering feet. Jack reports that Brian Beat told Stevie, "I was nobody's bitch," referring to Brian's eighteen years behind bars. So Beat was strong. He was dominant, he would like us to believe. Later, in the same conversation, Brian Beat told Stevie, "I was everybody's bitch." So Brian wants us to know, he was open to all comers. But that isn't the worst of it. Stevie tells my RA that Brian Beat has a daughter. Stevie knows about the child because he is best friends with the child's uncle.

This cannot possibly be true. Do rapists really have daughters?

An inert body requires the application of force to set it in motion; I remember this from elementary physics. Why do I seem to lurch between a numb, inert state, when there doesn't seem to be an "I" capable of picking up the phone, and a state of pure anxiety, when momentum is almost impossible to stop? The news about Brian Beat's daughter supplies the needed force. I do not want to meet the daughter. That would hurt me. I don't know why. And it might hurt her. But I want to meet her mother.

Now I find myself able to ask Chet if he can drive me out to Milbridge again. He agrees. He will take a day off from work, he says, to drive me there. We set a date to drive out to Global Gas to meet the uncle of this child, the child of the man whom the police believe raped me.

Yes, I can carry out this project, but my goodness, I need help.

Before we leave my apartment, I suddenly realize the laundry needs washing. I scrub the kitchen. I make the beds. When this is done, I begin organizing my files. I also change my clothing several times. I know that I need to wear a long black skirt, a white blouse, and a black jacket. A kind of puritan uniform. But I have two jackets. Which one? Eventually my frenzy gets on Chet's nerves. It is time to go.

It takes an hour to get to Milbridge. Chet and I don't talk much. I feel shaky. Perhaps the cleaning frenzy has enervated me.

Global Gas is on South Main Street, just past the police station, around the corner from the house where Brian Beat grew up. When we walk into the body shop, I have a kind of tunnel vision. I see Stevie, Brian Beat's friend, who seems to leer at me expectantly, and Uncle Henry, who looks pained.

Henry offers a gnarled, oil-stained hand to Chet, who shakes it manfully. My hands flutter, ashamed of themselves, to my sides. I am alarmed by the oil stains, and ashamed of my alarm. I am also deterred by the pain I sense in Henry's hands.

"I have arthritis," he says, relieving me, at least momentarily, of the misery I feel at my inability to touch him, or anyone else for that matter, at this moment.

"Could we go for coffee?" I ask.

I have an urgent need to get away from that bathroom.

"I don't have much time," Henry says.

"Someplace where we can sit," I say.

There is a Dunkin' Donuts about a mile away. We follow Henry's black pickup truck with its artificially engorged tires. I sense my features changing. My eyes seem to swell. I cannot focus.

Do rapists really have daughters?

A cage floats down from the heavens, encasing my body in glass. Chet does not see the cage. He thinks I am with him in the car, but in fact, I am alone in a parallel world. I look and sound relatively normal, but really, I'm only half here, only half alive.

We turn the car into the entrance of the coffee shop. The ticking of the turn signal scrapes against my inner ear.

Henry waits for us inside the coffee shop. We find an empty table. We sit under the nauseating glare of fluorescent lights. I hear the dizzyingly familiar buzz. Terrors are brought back to me by this sound, but I don't know what they are or why. The coffee creamer feels far away, but the ceiling presses close. The hand that holds my coffee cup does not look like mine. I sip my coffee carefully, worried that I might drop the cup or spill the hot liquid. There is a strangely altered distance between table and mouth. Here is what makes me feel so very alone in these moments: nobody else notices that I am no longer in the room.

I am not entirely sure that I will live through this interview, so the best thing is to jump right in, get it over with.

"Tell me about your sister and Brian Beat," I say to Uncle Henry.

"I don't know exactly what happened," he says. "When she got pregnant, I was in Vietnam. My sister, after she got pregnant, she moved to Connecticut. She married someone else. Beat went out to California. He was dating a girl out there. The girl's father was a cop. The father didn't like Beat. I think she accused Beat of rape."

So Brian Beat may have been accused of rape in California, too.

"I didn't like Beat," he confesses. "But I think he was falsely accused. I don't believe he raped anyone. He was smart. He was brilliant. He could have done anything.

"But he was leading my sister down the wrong road," he says. "They would run off to Webster or to Worcester. They were taking drugs, and I was the one sent off to find them.

"You really have to talk to my sisters," he says, anxious to be done with me.

"I would like to," I say, equally anxious to be done with this interview, mainly because I'm so dizzy. I ask Henry if he would call his sister Abby to tell her that I'd like to talk to her.

"I'll call Fay," he says, referring to their younger sister.

"Abby is difficult," he says. Henry and Abby are mostly out of touch. He does not even have her number.

I notice that the skin on his face looks raw, as if his flesh had been lashed by the wind.

"She's had a very hard life," he adds, defending his sister from some unspecified person's critical eye.

But then, as if to correct the record, he says, "She brought it all on herself."

I hear the sound of unrelenting rain. Such bleakness here in Milbridge.

He pulls himself up taller now, reining himself in. "She made her bed. Now she'll have to lie in it," he concludes.

I look around the coffee shop and then back at our table. Now I notice that it is dented and unsteady, that some of the silverware is bent from overuse. Scruffy white tufts of an asbestos-like substance are protruding from the back of Henry's chair, which is upholstered in gray Naugahyde, the same color as the examining table in my grandfather's office.

There is a silence. I wonder to myself, What bed does he have in mind? The one she made with Brian Beat nearly forty years ago? Such a common expression, I think to myself, this making of the bed that now we have to lie in. But so very cruel.

He must have sensed my confusion.

"She used to be a nurse, but she got into a fight with someone at work. The police were brought in to investigate. They found that she had been caught with a needle in her purse when she was fourteen years old. It was still on her record," he tells me, "from the time she was hanging out with Beat. They fired her. She's on disability now."

While he talks, he dials his sister Fay's number. She is at her desk at work. I see Henry's face soften as he hears his younger sister's voice. They chat briefly. Then he tells Fay that a researcher is interested in talking to Abby.

Henry hands me the phone. I tell Fay I'm conducting research on trauma and violence, and that I'm hoping to talk to Abby about Brian Beat. Fay doesn't seem entirely surprised, though I hear suspicion in her voice.

"A reporter came to talk to us once before," she scoffs. From the sound of her voice, she does not appear to have held this reporter in high esteem. "Sally, our oldest sister, was murdered," she says, "and a reporter came to investigate. But the story never ran."

I catch my breath.

"When?" I ask, wondering if Beat could have been the murderer.

"In 1974," she says. "We know who murdered Sally. It was her husband. We know it. But he was friends with all the cops up there. Up in Gorham. Friends with the DA. There was a big cover-up," she spits.

I ask more questions, half horrified, half incredulous. How could there be so much violence hatched in this tiny town? Later, I will find press coverage of Sally's murder, which occurred in August 1974, several months after Brian Beat was imprisoned. So he wasn't a murderer. At least, he didn't murder Sally.

Now that she has this story about the unreliable reporter off her chest, she is ready to give me her sister's number. But first she wants to tell me her own view of Brian Beat.

"Everyone was shocked when Brian was arrested," she says. "He was just gorgeous," she adds, breathlessly, as if Brian's "gorgeousness" were clear proof of his innocence, as if she is confident that this one word will dispel any doubts I might harbor. There is a girlishness to her voice now, a gossipy tone. She seems to relish the memory of her big sister's beau. "The girls were throwing themselves at him. I don't believe he could possibly have raped anyone; he wouldn't need to," she says. "He dressed—well, you would be amazed. Like someone, what do you call that magazine, like someone out of *GQ*. But you really need to speak to Abby—she is the mother of his child."

Finally, she gives me Abby's number.

I am still sitting across from Henry under the buzz and glare of the fluorescent lights. The perspective in the room is still annoyingly off.

I dial Abby's number. She does not answer. In my daze, I let the phone keep ringing. A ringing in my head. Finally, there is a woman's voice.

"Hello," I hear. A kind of croak.

I explain that I am researching cycles of trauma and violence and that I am hoping to speak with her about Brian.

She takes in this information. She does not appear to be shocked. Perhaps she, too, talked with the reporter after Sally's death. But she says, "I'm not well." There is a pause. "I'm not even dressed."

Perhaps she senses disappointment in my silence. "I will be willing to talk to you about this if you can drive down another time," she says.

I am devastated. Childishly so. In this moment, I am not at all sure I will ever be able to come back to this part of the state. But I manage to recover a sense of professional equanimity. This is not the first time a potential interviewee has refused to talk to me. Once I flew all the way to Amman to talk to Hamas members there, only to be rebuffed when I arrived.

Suddenly, for no apparent reason, she changes her mind. Perhaps she is curious.

"My house is a mess, so we'll have to go out," she warns me. "Give me an hour to get dressed," she adds. I look at my watch. It is 12:45 PM.

When we take our leave of Henry, I have trouble locating the car in the small lot behind the coffee shop. The car looks different—longer, or maybe shorter. I feel that I am different, too. My features feel even more swollen. The phrase *features swollen like those of a mongoloid child* come to my mind. That was the cruel term used for hydrocephalus when I was a child.

"Do I look different?" I ask Chet.

He tells me that something may have happened to my pupils.

We drive the fifteen miles to Worcester, then stop for lunch to allow Abby time to dress. Chet, as usual, is in a mood to explore. He is delighted to be driving to a part of Worcester he's never seen, delighted to try a new restaurant, delighted to meet someone new. I am not at all delighted by any of this. I have an overwhelming desire to shut myself in like a bivalve. But another feeling competes with my desire to hide—I am curious. What will I learn from Abby? And I also want to know, what will happen to my body and mind when I hear whatever it is she has to say?

After lunch, we make our way to Abby's house. She has provided directions, but she hasn't taken one-way streets into account. After several false turns, and several stops to consult passersby, we finally find Abby's street, and then her home. She lives in an apartment in a triple-decker house. The house has gray siding. The siding is slipping.

I call from the street to tell Abby we're outside. She comes downstairs. She locates our car. It is still raining, with a heavy gray sky, but she blinks, as if adjusting to the light of day. I notice long straight hair, long legs, blue jeans, a 1960s look. Her face is traced with delicate lines. She has no scent. Remnants of her good looks remain, intensifying an overwhelming sense of loss I feel in her presence. The vulnerability in her glittery blue eyes makes me feel I ought to look away. It is as if, in observing the lines that faintly crease her delicate skin, I am trespassing on some private grief.

"Dunkin' Donuts is just a couple miles down the road," she says.

I offer her the backseat, then scramble to gather my son's leavings—half-eaten muffins, mittens, an extra pair of boots. She sits down and buckles herself in, but then suddenly comes to her senses.

"Can I see some ID or something?" she demands. "How do I know who you are?"

I assume that she is concerned about getting into a stranger's car for all the normal reasons. We might be rapists, or even terrorists. Okay, those might not be totally normal worries, but you never know. But she is not worried about her own safety, it turns out.

"How do I know you're not working for the feds?" she asks. "Brian is dead, but I still don't want to hurt him in any way."

I paw through the papers stuffed carelessly in my purse. Why am I not more organized? Old receipts. Library notices. The results of medical tests. Other people's business cards, but alas, none of my own. Panicked, I offer up a driver's license and a Harvard ID. For whatever reason, these two items help to persuade her that I'm not a fed. She agrees to proceed.

She directs us to a shopping mall a few minutes down the road. Dunkin' Donuts is on the corner, just outside the mall. When we arrive, the coffee shop is crowded with retirees enjoying a mid-afternoon snack. I am worried that the happy din will be too much for Abby, but she seems unfazed. Chet asks Abby what she would like to eat. She requests coffee and a chocolate cruller. I wonder if this is her first meal of the day.

I ask her how she first met Brian.

"I was fourteen or fifteen," she says. "He was a couple of years older. He had been going to a vocational school, an agricultural school. When I met him, he had just switched over to Milbridge High. There were three of us who started hanging out together:

Brian, me, and Simon Brown. Later, John Henry was part of our group. We were oddballs. People called us hippies.

"He was very intelligent. He knew everything about everything, things that a lot of us weren't interested in or didn't know anything about—politics, history. He was spoiled by his parents. His mother—she adopted him, she wasn't able to have children—she just adored him. He was her only child."

Now I recall the document that Lt. Macone gave me, which indicated that Brian was adopted.

I ask about his birth mother, the sister of his adoptive mother. "He never got over the fact that his birth mother gave him up. She wasn't interested in him. She went on to have more children, but that side of his family only took an interest in Brian after his adoptive mother died, when it came to settling the estate.

"He hid a lot of things, the pain he felt at not being wanted by his birth mother. He told me things he wouldn't tell anyone else.

"He was so gorgeous," she says, in agreement with her younger sister, at least on the matter of the gorgeousness of Brian Beat, a quality not at all consistent with my vague recollection of a skinny man with a gun, and not at all in evidence in the photographs I now have of him from his prison file, the photographs I barely caught a glimpse of, but an impression remains nonetheless.

"He went for other girls, too. I know a lot of girls were after him. He needed to have conquests. But I always felt that he did that because he wanted to prove he was a man.

"I was thin," she adds. "One hundred and fifteen pounds, but he thought I was heavy. He thought I was supposed to look like Twiggy. I'm five foot five, that is my normal weight. But he made me feel like I was fat."

"How old were you when you got pregnant?" I ask.

"Seventeen," she says. "In my senior year of high school. We didn't have birth control back then. Brian wanted me to have an abortion, but I refused."

Abby's father was Brian Beat's father's best friend. The two fathers knew about Abby's pregnancy. But Mr. Beat supposedly never informed his wife that their son had impregnated Abby.

"My parents wanted to put me in one of those homes for unwed mothers. Girls who go there agree to give up the baby. I didn't want to do that. So my mother called my sister Sally, and I moved in with her."

She lived with her sister Sally for a while, and ended up marrying the brother of Sally's husband. The brother of the man whom Abby and her sister Fay believe murdered their sister Sally. How could one small town sustain this much violence?

"I married Tom when I was eight months pregnant. He didn't care that I was pregnant with another man's child. He said it is one thing to impregnate a woman, another thing to father a child. He never threw anything into my face. He was very good to me. He loved my daughter. I didn't have any contact with Brian when I was married to Tom, living in Connecticut.

"But I was in love with Brian," she says. "I really wanted to be with him." She is smiling now, for the first time, drifting into a wistful space. The horror of it shocks me awake.

But her wistful mood quickly fades. "None of my relationships since then have worked out, because I was really in love with Brian," she says.

Is it possible to be in love with a child rapist?

"He had a really hard time in prison. The other inmates mistreated him. They set fire to his bunk. They thought he was a diddler."

I ask what that means. "A child molester," she says. "They don't like diddlers in prison," she explains.

"When he came out on furlough, he kept saying to me, 'I didn't do it.'

"After he got out of prison, we were thinking about getting back together," she continues. "But he was unstable. He wrote me a letter about how proud he was of me that I kept my daughter. But he kept asking, Are you sure it's mine? That really hurt me. There was a cold side to him. I realize now, looking back, that he was very cold when he made love to me. There was a lack of compassion during the sex act."

"Do you think he could have been abused?" I ask.

"Not by his parents," she says. "Maybe by someone else. There was a lot of sexual abuse and incest in Milbridge," she adds. "No one was ever arrested, but the police knew about it. It was written up in an almanac, an annual report on the town. I was shocked when I saw it. There were over a hundred reports of incest, women getting abused, domestic violence. The police never did anything. Back then, they assumed that if a woman was getting beaten, she must have done something wrong.

"I really liked his mother. She told me that she always thought the two of us would get together. But she told me, You're too good for him. She knew there was something not right with Brian's mind—but people didn't talk about these things in those days. He was moody. Sometimes he was extremely charismatic. He kept himself really well. He always had to have his tan. He wore a corduroy suit. . . . People were intimidated by him. He was so bright. But sometimes he could be very cold. He had different personalities. He could be very charismatic and loving, but he could also be very cold. Not a lot of compassion.

"Karen was around thirteen when I told her that Brian was her father. She's been mad at me ever since," she says.

I don't want to interrupt her story to ask who Karen is; I assume that she must be Abby's daughter.

But Abby suddenly recollects that she's not talking to an old friend. "My daughter," she explains.

I do not ask where this daughter is. I do not want to meet her.

"Should I have told her or not?" she asks.

I am relieved that she doesn't wait for my response.

"I would have wanted to be told who my father was, even if he were accused of rape. I did what I thought was right. He stopped over a couple of times. She was a little afraid of him. Sometimes he acted bizarrely. I wanted to protect her."

Protect her from what, exactly? I cannot follow the flow of her logic here, but again, I don't want to interrupt her.

"Things could have happened for him. He was so smart. But I just couldn't save him. I had to protect my daughter," she repeats.

If she is persuaded that Brian Beat was innocent of raping children, from what, exactly, was she feeling the need to protect her daughter?

After he got out of prison, he changed, she repeats. "He became a loner, a recluse. He walked all day long, from Milbridge, to Webster, and back. Once I picked him up; I'm not sure he even recognized me. But he was still gorgeous. He was so strong. He walked all the time, very muscular. He was in fantastic shape.

"It was so awful—Why did he hang himself? He had nothing. Sometimes you think about these things. Someone I cared so much about. But I just couldn't take him in. He was crazy. I was a nurse. I know he should have been on meds. You can't help someone like that if they won't take their meds."

She is crying now, sobbing, actually. I consider whether I ought to hug her, but decide that it would be dishonest. How can I comfort her about the death of the man the police believe raped my sister and me? I feel an overwhelming sense of compassion toward Abby, but I am immobilized by confusion. I can't comfort her—not only because she is mourning the death of a man who was convicted of raping children, but also because I don't know how.

In this moment I tell myself that she and I are of different species: she—outwardly tough but inwardly shattered; I—effete on the outside but tough as nails on the inside. She looks so alone, so vulnerable, sitting there on a cold metal chair under the stark lighting, surrounded by her neighbors whom she doesn't know. But the compassion I feel is mixed with horror, shame, moralistic judgment, and guilt. In her presence I am profoundly aware of how unfair life is. She grew up in Milbridge, Massachusetts—where drunks and pedophiles were common and commonly on display, and where girls learned to expect abuse and prepare for it. I grew up in Concord—where drunks and pedophiles are well-bred and secretive, where good girls, of which, at this moment, I know I am not one, learn the fine art of denial.

She pulls herself together.

Out of the blue, she tells me, "I saw my father sexually abusing my older sister. I was so afraid. We were moving into a new house, and we were staying in an apartment. The apartment had one large room. Sally was in a bed in the corner, and my sister Fay and I were sleeping in a double bed. He came in, in the middle of the night. He was half shit-faced, as we used to say."

This is just impossible, I'm getting more than I asked for. Can I ask her to stop, to keep these secrets to herself? I know that I cannot. It would be cruel.

"He went right over to Sally's bed. I covered my face. I was young; I didn't really understand what was going on. I was young, but I wasn't stupid. She didn't yell, but I know I didn't imagine it. After that I never trusted my father. I always left my door open a particular way, so that if he opened it to come in, it would squeak.

"Even back then I didn't sleep that well," she says, launching into a discussion of the effects of a medication she takes for sleep.

"Where was your mother?" I ask.

"She had gone to the hospital for a gallbladder operation," Abby explains. "She left us alone with him."

I know all about this, these mothers who get sick and the wolves who move in, but I remain silent.

As I leave, I spell out my name for Abby and give her my phone number. "Jessica Stern," she intones slowly. "That name will be famous someday." Not, "You will be famous someday," but "that name." I ask what she means by that, thinking that perhaps she has seen my name in connection with my work on terrorism. "I just think we will all know that name someday," she says. Does she mean that my name will mean something to her in the future that it doesn't mean to her today?

I leave Abby, strongly considering the possibility that Brian Beat was not my rapist. No, that's not quite right. Let me try again. I leave Abby, doubting that I was raped at all. Abby's face and Abby's world seem much more real to me now than my own rape. And I am suddenly worried about hurting Abby, or especially her daughter.

We are sisters of sorts, both having had sex with this man. But the man she had sex with was "gorgeous" and "charismatic," even if "cold during the sex act." The man I had sex with—if we can call rape sex— Let me start again. The man who penetrated me was a skinny pedophile for whom foreplay included demanding, under threat of death, that my sister and I put on our little sisters' clothing. Our sisters were then eight and nine. That man penetrated me with his shame.

Shame, I realize now, is an infectious disease. Shame can be sexually transmitted.

Now that I have learned that Simon Brown was another member of Brian Beat's close circle of friends, I do what I can to find him. Once again, I find it hard to call anyone who knew Brian Beat.

Once again, Jack locates Simon, and even makes an appointment for me.

Simon has asked me to come to his office, which is located in an old industrial city near my rapist's hometown. He runs a company that supplies sprinklers for use in large buildings. Chet has taken the day off, once again, to drive me to my interview.

The closer we get to Simon Brown's office, the more detached I feel, as if I were floating slightly above the ground. In the car, I cannot bear the sound of music. The notes are sharp, or flat, or in any case not right.

The building where his office is housed is stark and anesthetized-looking. Once inside the building, under the fluorescent lights, I feel even worse. I lose my ground in generic buildings like this one, the sort where inside, you could be anywhere in America. Without a sense of place, I float even higher. But you probably wouldn't notice that I'm floating if you saw me in this state. I would seem more officious than normal, more efficient, perhaps a bit rushed and cold; but not like a person who doesn't sense the ground under her feet.

Simon Brown's receptionist pushes a buzzer and informs him over an intercom, "Your ten o'clock is here."

Simon Brown is ready for me. The assistant directs me to his office, right around the corner from where she sits.

I try to observe him carefully. It takes an effort to force my brain to process visual stimuli in this floaty state. I'm far more likely to notice sounds or moods or scents. I note that he is slender, with gray hair. He has handsome features, almost stately. He looks respectable. I sense that fastidiousness and debauchery might have wrestled for control at some point in the past. Fastidiousness won the battle long ago.

My attention is immediately drawn to the sound from a printer in the corner, which is spitting out blueprints. I ask him what the plans are for. "It's a blueprint for a very large building. We do everything on CAD [computer-aided design] now," he tells me.

I am relieved that there is something being designed and produced here. There is something for my mind to catch hold of. Details of other people's lives always tether me. To have one's attention held—to be fully engaged in anything—relieves this floaty feeling, which is painfully annoying. It is like being stuck between earth and heaven, not quite alive but also not dead, accessible to neither people nor angels.

Simon answers my questions about how long he's been in business and who his clients are, and asks if I know a government-office building in Central Square where he has a new contract requiring his staff to acquire security clearances.

Eventually we turn to the business at hand. He has heard from my research assistant that I am writing a book about victims and perpetrators that is partly about Brian Beat.

Without being asked, he tells me right away, "I don't believe Beat raped anyone. I'm quite certain of it," he adds.

"Why are you so certain?" I ask.

"I don't want to go into all the details," he says, apparently imagining that the intimate details of his life or the life of his dead friend would be boring to a stranger. I want to urge him on; I am never bored by people's life stories, or by the stories they tell themselves. But I restrain myself, confining myself to one question.

"Was Brian Beat gay?" I ask, puzzled by Simon's caginess.

"No, I'm sure he wasn't gay." Again, Simon doesn't hesitate.

"Well—I can't speak for what happened when he was in prison. But he wasn't gay when I knew him," Simon adds, unnecessarily.

Once again, I am aware of the strong planes of Simon's face, as if an architect had a hand in designing his features. The planes have softened with age, and there is that whiff of sensuality about his mouth, apparently renounced in favor of propriety. In this altered state, which I've been in before, I can feel my way—as if my mind had fingers, as if his features were a form of braille—into what I imagine is my interlocutor's character.

Oddly enough, when I am in this altered state, my interlocutors bare their souls. The communication between us is a life raft in a wild river, as if both of us would drown if we didn't speak. My job is to ask, their job to tell. It feels to me, in interviews like this one, that if I don't ask (with or without words), if they don't answer, we will drag each other down in a sea of fear. I cannot control whether I enter this altered state, but it almost always happens when I'm interviewing terrorists.

Here I see propriety softened by the sadness of unmet expectations. In place of those expectations I sense kindness, acceptance. There is a hint of shame in the corners of his mouth, which you might not notice right away, distracted by the handsome planes of his cheeks. I note these fleeting observations, and I place them aside. Even though I've been through this before, at some level I know that these judgments might reflect my own prejudices. I will let Simon's story unfold in his words, at his pace.

"I was living out of state at the time he was accused. But I have a hard time imagining . . . We dated the same girls. A lot of them. It was the period of, you know, sex, drugs, and rock and roll. Not necessarily in that order.

"I've had no contact with him to speak of since I was twenty-one," he tells me, "other than seeing him on the street."

"You mean when he was a street person?" I ask.

"Yes," he confirms, his eyes looking out now at some horizon, some sad memory.

"You never met Brian?" he asks.

"No," I say.

In this moment I have fully inhabited his sadness, the sadness of observing a close friend throw his life away.

A minute later I am shocked to realize what I have done: I just lied. Worse still, I fully believed the lie in the moment I uttered it. I am a reporter here, not a victim of rape, certainly not the victim of this Brian we're discussing. Now I begin to watch myself more closely, wanting to avoid inadvertently lying again, wanting to avoid inadvertently revealing more than I intend. I will correct this lie, I tell myself, at the first opportunity.

"How old were you when you met him?" I ask.

"We moved to Milbridge when I was in third grade. We were very good friends from the time we were eight years old. We went to school together until high school. I went to Milbridge High. He went to high school in Worcester, to an agricultural school. A surprising choice for him. He was extremely bright, extremely entrepreneurial," Simon explains.

"What do you mean by entrepreneurial?" I ask.

"You never met him . . . ," he says. And I interject. "Actually, I did. I completely forgot about it. I met him, very briefly, a long time ago."

On a track of his own, he lets this go by, as if it didn't matter. Or maybe he already knows the actual source of my curiosity about his childhood friend. Perhaps it is easier to imagine that I didn't know Brian. In any case, I didn't know the Brian he is describing.

He continues. "One summer we bought root beer and sold it on the street. We each made around a thousand dollars. Another time we bought several hundred ears of corn, and we walked door to door to sell it. We made a hundred to two hundred dollars per week.

"He was a risk taker. Very curious. Always wanting to try something new. He would often get into quote/unquote 'trouble.' There was a river near our houses. We trapped muskrats to sell the fur. When the ice starting breaking up, we would use icebergs as a boats, travel down the river that way."

"Do you think he might have been sexually abused by a priest?" I ask.

Simon doesn't seem surprised by the question, and he has a ready answer.

"No. It's the sort of thing he would have told me about. I'm sure he would have told me if he were," he says.

I wonder to myself, How can he be so confident?

"And anyway, he went to church in grammar school and middle school, but he stopped going by the time he got to high school. His mother wanted him to go, but he always found a way out of it."

"It might have happened when he was really young," I suggest. "When he was at that Catholic elementary school. And it isn't the sort of thing that kids necessarily talked about back then. You would have had to be really close. Were you really that close?" I ask.

I have a slightly heady feeling—as if we have switched to a different atmospheric plane, where the normal social rules don't apply. The way you might ask impertinent but crucial questions when you are speaking to someone who is about to die. Will he be annoyed?

He seems surprised rather than annoyed, as if he had never thought about this, but also persuaded. "I guess you're right," he says. "But I still think I would have known if the priest at Saint Roch's was abusing boys. Everyone in the neighborhood was Catholic. And I used to see the priest at social functions, at dances."

I don't bother telling him how many of the priests at Saint Roch's were kicked out for preying on young children, or the rumors about sexual abuse at Brian's school.

Time to change the topic.

"People keep telling me that he changed when he realized he was adopted. Did you notice that?" I ask.

He agrees. "What he told me was that his biological mother had an affair when she was in high school. His biological mother gave him to her older sister to raise. He would have been approaching sixty today; he was born in '47. In those days it was unheard-of for a woman to raise a child on her own. He was a pretty happy-go-lucky kid—and then he discovered that his cousin was actually his sister, his aunt was actually his mother."

"Did his biological mother ever marry?" I ask.

"No. Filomena was a little nutty. She had a long-term boyfriend. But twice a year she would break up with him, and she would take up with two or three other men. Then she would get back together with her boyfriend. It was very strange for us, to see Brian's aunt, who was actually his mother, showing up at his house with different men. We lived in a stable neighborhood. Everything was very hush-hush. So this was pretty shocking."

"How old was Brian's birth mother when she had the daughter she kept?" I ask.

"Carla was two years younger than Brian. Only two years younger," he repeats, as if for the first time noticing how painful it must have been for Brian to learn that his birth mother felt able to raise his sister on her own, but not able to raise him, even though she still wasn't married when she gave birth to his younger sister two years after she gave him up for adoption.

"I saw Filomena's daughter at Brian's funeral. Carla. I liked Carla," he says.

"Were Carla and Brian close?" I ask.

"They were quite close when they thought they were cousins. And they stayed close when they learned they were brother and sister. We saw a lot of them. Filomena came by around twice a month. I liked Carla," he repeats.

"I heard that there was a group of you that hung around together. Brian, you, Abby, John. Is that right?"

"Yes," he confirms. "And Carla," he adds.

"Did you hear that John Henry had died?" I ask.

"It wouldn't surprise me. He took a lot of drugs. We all did back then, but Brian and John—they didn't stop.

"Brian and I went to jail together because of drugs," he offers. Some painful truths are fair game, it seems.

Out of the corner of my eye I observe a new awkwardness in his demeanor, a look of distaste, as if the proper side of Simon were half disgusted with the more sensual side he left behind. The proper side was still in control, but was prepared to allow the hungrier side to unburden itself of some shame, to seek absolution.

"We were in Bridgewater State together for a couple of months. Then in Worcester. After that I was done with that life. No more drugs.

"And I was finished with Brian, in fact," he says, as if washing his hands of Brian yet again.

"Sex, drugs, and rock and roll," he repeats. "I'm not at all proud of it. It was a time of experimentation. We all experimented. But jail hit me pretty hard. It wasn't just a slap on the wrist.

"I've never told my son about this," he confides. "He knows I didn't go to college, but he has no idea why."

I am intensely interested in this topic—the secrets that don't get told, that become malignant in the not-telling. But I stop myself from asking more. Brian is the story here.

"Did Brian get raped in jail?" I ask.

"No. He didn't," Simon says. Too quickly. I wonder to myself, once again, why he is so certain, why he is so sure he would have known, but I let the subject drop.

"Abby says that Brian was very good-looking. Is that true?" I ask. "What did he look like?"

"He was very good-looking. Same physical size as me. He had good features. Blond hair—a Beach Boy look."

"Abby says that she was really in love with him. Do you think that is true?" I ask.

"We both had girlfriends. . . . You were of the moment back then. She was infatuated with him. She was in love with him as much as a sixteen-year-old is capable of being in love. But it wasn't reciprocated, I can tell you that. She was available. And other girls were also available."

"Did you ever see any signs of instability in Brian?" I ask.

"No. He was pretty rational."

"Apparently he was obsessed with poetry. Did you ever notice that?"

"I was not aware of any obsession," he says. "But he wrote me a couple of letters from prison that had some poetry in them." He lost the letters, he tells me, when he got divorced and moved.

"When we were in high school, Brian and I were on Highland Street, and some kids from Cambridge came by. We went to a party with them. We ended up doing acid with Timothy Leary," he says.

"Were you part of an experiment?" I ask.

"No! It was just a party. His use of acid was not strictly academic." He corrects me, somewhat prissily. I imagine he is nursing a judgment against academics—their naïveté, their veinless pedantry, so removed from real lives and real passions.

"What was he like?" I ask.

"I don't know. I was stoned out of my mind. I didn't realize who he was until later, when I read about him in a magazine.

"You never knew where we'd end up. Friday or Saturday night, someone would say, There's a party over there, in whatever town, and we'd go.

"Brian was the one with a car. Filomena gave him a car at one point. When he was sixteen. For his birthday, or maybe it was Christmas. She gave him what seemed to us very extravagant gifts."

"What kind of car was it?" I ask.

"A Studebaker. They don't make them anymore. A green Studebaker.

"It wasn't new," he adds. "But still it seemed an extravagant gift to us at the time."

"Did he seem tortured?" I ask, wanting to avoid an extended discussion on the topic of cars.

"No. Maybe about his father," he reconsiders. "He felt really bad that his father gave him up, didn't want him."

"What about his adoptive father?" I ask. "What was he like?"

"Ken was fairly strict. But Ken and Ellen—they cared for him a lot. Until the drugs kicked in. Then they really didn't know how to handle him. They would try being strict. They would ground him. But he would sneak out of the house after they went to bed. Then they tried being lenient, but that didn't work either.

"He might have had signs of—what do you call that?—OCD," he says, apparently reconsidering his earlier position.

"What do you mean?" I ask.

"He would insist on listening to the Beach Boys twenty-four/seven."

The 1990s expression is jarring. I'm feeling the mood of the 1960s—before hard work and commercialism became so fashionable. I have entered this story and entered Simon's mood; a habit I learned trying to get terrorists to talk to me, a style of conversation that comes naturally to me now when I'm trying to understand something that terrifies me. But as Simon continues, I'm pulled right back in.

"Then it would be Dylan. Then it would be the Stones. If someone tried to stop him, he would get angry. There was no moderation. He was gung ho.

"And he was a risk taker," Simon repeats. "He had the money to buy gas for his car, but he always wanted to steal gas from a farmer down the road instead of buying it because he thought stealing was more fun. He got a thrill out of doing these things."

"Do you know how he lost his virginity?" I ask, aware that I'm intruding onto territory where I don't belong. Not because we're talking about the man who I suspect was my rapist. At this moment we aren't—we are talking about Simon's childhood friend and, I suspect, his lover (perhaps the lover that Brian Beat referred to in his efforts to get out of the draft). I am feeling myself to be a teenager at this moment, not cool enough to be part of their clique. Not the sort of girl who would have been friendly with these boys, but the sort they would disdain. And I'm poaching, pretending to myself and to Simon that I would have been their kind.

"We were both twelve," he says. "We talked about it. We wanted to do it. There was a girl in our neighborhood. She was fifteen. She was developed. She looked good in a bathing suit. She was willing to have sex with us. So we all did it, together," he states—as if wanting a girl and having a girl were tantamount to the same thing.

"You mean you were together in the same room?" I ask, surprised.

"Yes," he confirms.

"Did Brian like it?" I ask.

I cannot imagine where I got the audacity to ask these questions or why Simon didn't kick me right out of his office. There must be a part of me that needs to imagine my rapist as a virgin—in order to see him as a human being rather than a beast, an aberration. And there must be a part of Simon that knows I need this and wants me to have it.

"I think so," Simon says.

"Did you remain friends with her?" I ask.

"Yes," he says. "She was one of the girls we had sex with. We also had sex with her younger sister. She was twelve or thirteen."

So she was fifteen, and had a younger sister, too. I don't process this right away. All I notice is that something about this story makes me fly higher—my self now out of the room, out of reach.

And then it comes to me. I was fifteen, too. I, too, had a little sister. The forcing that ripped my flesh, tore it so that it would never return to its original state. My younger sister was fourteen. The two girls. The watching. The theater.

"After we got out of prison, I said to Brian, Brian, this is not for me. I sort of cut off our relationship then. We were seventeen or eighteen. I would run into his parents, and they would tell me what he was up to. He and John Henry went down to the Cape. They worked down there for a while."

I ask Simon if there is anything else he thinks I ought to know about this person, Brian Beat, victim, perpetrator.

He thinks for a minute, and then repeats, apologetically, "I really cut off relations with him when we were seventeen or eighteen, so there really isn't much else I can add."

For reasons I can't really explain, I blurt out, "You really should tell your son about the period you spent in prison. I think it would help him to know this."

He seems to consider the idea.

Why did I say that? I am embarrassed by my outburst.

Still, I have an idea that both of us have been changed, at least in a small way, by what we shared in the last hour. And that he will probably tell his son soon, and that his son will be relieved to hear the story. And that the planes of his face will settle into a less shamed state.

Jack, my intrepid research assistant, has gone back to Milbridge several times, to see what he can learn by hanging around at Global Gas. He discovers that one of the Beats' neighbors, who lived there at the time Brian Beat was growing up, is still living in the same house.

I call Mary several times, and eventually she agrees to meet with me. Chet, once again, agrees to drive me. Once again, he takes time off from work. I'm terribly embarrassed about having to ask him, but we both know that I won't make it if I drive alone.

We chat cheerfully in the car, as if we were on an outing. Chet knows the way now, and I have become an old hand at being driven out to this awful town. It is hot. As we get closer, I begin to sink. I have no feeling I can name. Only a sense of wilting. I am lost in a sea of sleepiness.

The houses in the area seem tiny to a twenty-first-century eye. A post–World War II development, built for millworkers seeking a new life in the suburbs. Were people smaller then? I feel claustrophobic even looking at these houses, even more so when Mary opens the door to her parlor. She is a small person. She just got a perm, right before we arrived. Her hair is done up in a pale blue bouffant. A suffocating profusion of drapes, chests, and glass display cases crowds out light and air. "Sit down." She smiles, pointing to a velvet sofa, robin's-egg blue. I sink down into spent foam depths. She offers me a glass of water. I am so hot after the long drive. There are silver- and gold-colored coins painted on the sides of the glass. An embarrassing string of unkind thoughts come into my mind: contaminated wells, lead paint, poverty, incest, mentally retarded children.

I drink the water. The side table is crowded with photographs and statuettes. Kittens and children. I'm confused about where to place the glass. I push a kitten to the side. I feel awkward, too big for this house, too big for Mary.

I know that Brian Beat lived in this neighborhood, but until Mary tells me, I don't realize he grew up right across the street. "They were here when we moved in," Mary tells me, pointing to a small house directly across the street. "They were the original owners of that house."

I see a coffin-shaped house across the street.

"When was it built?" I ask. Nineteen forty-nine, she tells me. All the houses on the street were built back then.

"How well did you know Brian?" I ask.

"I watched him grow up," she says. "He was a very polite kid.

"Disturbed," she inserts. I check my notes several times. Could I have imagined that she said that?

"I was divorced, and I had to work. I used to drop my daughter off at their house to wait for the bus there.

"He hung himself in the garage," she says. "Right by the door that goes into the kitchen. I had gone out bowling that day. Right there, in that garage."

She points across the street, to a white garage. The houses here, everything so neat and small.

"At one point I had a serious argument with her," Mary says, referring to Brian Beat's mother.

"What was the argument about?" I ask.

"I can't tell you. I don't like to speak ill of people. I don't like to speak ill of the dead," she says.

As she says this, she pulls herself up to her full height. She must be four foot ten. I feel myself cowering slightly under her gaze, like a child who has been reprimanded for sneaking around, spying on people. I am not of your kind, the kind who would speak ill of the dead, she seems to be telling me.

I try to control my thoughts, but I discover that I cannot. My mind keeps returning to the image of Brian Beat, swinging from the door between the kitchen and the garage, in the small house across the narrow street.

"Tell me more about what Brian was like as a kid," I say, steering us, I hope, back to a more innocent time.

"He was a loner. He always dressed well. He was a couple of years old when they moved here. He was the same age as my daughter.

"He went to Catholic school. St. Louis. We aren't Catholic. But he didn't take to it, and he switched."

"Do you know why?" I ask.

"No, I don't know," she says. "I'm not Catholic," she repeats.

"Do you think it is possible that Brian was abused by a priest?"

I ask, astonishing myself with this question. Detective Remas was the one who put this thought in my mind, but it seems rude to mention this topic to an elderly lady.

"I wouldn't know," she says.

She thinks for a moment and then adds, "There are a lot of boys today who have been abused; it happens all the time around here. There was that priest in Milbridge who was shipped to a neighboring town. One in Northbridge, too. One in Webster. It was going on all over the place around here until someone was finally brave enough to speak up."

Now she drifts back to a later period, after Brian was released from prison. Did something happen back then that makes it hard for her to remain on the topic of the long-ago past? Something that made her mad at Brian Beat's mother?

"He walked a lot. He rode a bicycle. But he got into an accident, and the bike got ruined. When Kenell [Brian's father] had to go for treatments, Brian drove him. Every day he drove him. The treatments were in Worcester, off of 290.

"My daughter Kathy was good friends with his cousin Carla. Carla stayed here a lot. In the summer. Ellen [Brian's mother] took care of her a lot of the time."

"Ellen was a very critical person. She was blunt. Kenell was quiet. She would criticize me for anything," she adds, gratuitously. I can see that something remains unresolved between Mary and her former neighbor, even though Ellen is now dead.

"I still don't know why he committed suicide. It was a shame," she says, referring again to the recent past.

"He knew that she was going to visit Kenell's grave," she adds, taking Ellen's side for a moment, thinking about what it must have been like for Ellen to get up from breakfast and find her fifty-nine-year-old son hanging from the door frame, blocking her way to the car she intended to drive to the grave of her husband, whom she had buried less than three weeks before.

I am no longer conducting this interview. I let Mary talk and take notes, trying to follow the broken thread of her logic, while concentrating on avoiding thoughts of rape/rope.

"When he was going to high school, he quit several times. Moved down to the Cape. He would go to the Cape with only a nickel in his pocket. To Provincetown."

I look up from my notes to ask why he was attracted to Provincetown. She doesn't know.

How could a young man survive in Provincetown with only a nickel in his pocket?

"I went to the funeral. I was asked, so I went. The service was at the cemetery. It wasn't in the church. North cemetery. Near the high school. Past the center of town.

"Ellen never went to church. Kenell went every week. He went alone. Ellen was bossy. Anything she wanted him to do, he'd do. Caspar Milquetoast. She was a very jealous person. He couldn't do anything without her okay.

"I don't think anyone in Milbridge knew he'd been in jail," she says, lurching back to the topic of Brian Beat. This is not the impression I got from the police or from other people in Milbridge.

"But anyway, he was never on the list of people to be afraid of." Does she mean a list of sexual predators? Does she know that when the police realized that he had slipped through the cracks and were about to put him on that list, Brian hung himself?

"He was a nice-looking kid," she emphasizes. Again, the broken thread of logic. "He had everything to live for. They bought him whatever he wanted. But he changed when he knew he was adopted."

"Do you know how old Brian was when he found out he was adopted?" I ask.

"Well, it happened in grammar school. Ellen told me. On the playground. Some kid. You know how kids are. They got into an argument. The kid said, 'You haven't even got a father. You're adopted.' That is how he found out about it."

"How did that child know?" I ask.

"From his parents."

"Was Brian upset?"

"He got upset about it when he was a teenager. He was an only child. He never had anybody to talk to."

There was a pause.

I ask whether Kathy, Mary's daughter, might be willing to talk to me.

"No," she says, a little too loudly, slamming that door shut. "She doesn't want to think about any of this. I'm certain she would not."

Afterward, I wonder what memory Mary is protecting her daughter from.

I need to take my leave now. I can no longer keep my focus. The effort not to think about Brian Beat swinging across the street is pressing in on me.

I ask to use the bathroom. The toilet seat is made of that soft plastic that your buttocks sink into. Enveloping and warm, just like the sofa. I wash my hands with Mary's antibacterial soap. My hands feel stripped. Superficially clean. I smell chemically sweet, of generic ersatz flowers, the scent of a funeral home. I think of antibiotic-resistant bacteria. Howard Hughes, I tell myself. He would have been paranoid about this soap, too.

I take my leave. As we're walking out the door, Mary tells Chet, "He was a smart kid. But what he did was dumb."

What was dumb? The rape? The rope he hung himself from? Whatever he did or did not do to her daughter? Why does she tell Chet this, and not me?

• • •

We are outside at last. You can breathe out here. Chet proposes to drive me around the block. We find the stream where Simon Brown told us that Brian Beat played as a child; and where, later, when he was homeless, he lived. The water is barely running now, in this heat. The smell of skunk cabbage. Empty soda cans littering the side. I imagine Brian Beat's feet sinking into the black mud. I see him washing his clothing in the stream, hanging it to dry from sticks. Those dirty feet. I push the feet out of my mind.

We circle back to the Beats' house, 1 Orton Place. Built in 1949. A white fence encircles the yard. A small statue of Mother Mary. An angel. Two Norway spruces. Everything proper. Everything fenced in. We are in the sad hour of the afternoon, when the searing pain of the suburbs sinks into your heart.

We drive to the end of the dead-end road. The stream is wilder here, the water flows. The birds seem happy this year, I've noticed. Their song is not sweet; not today, at least. But it is joyful. As if they don't know a rapist of children lived and died on their street.

chapter six

Climbing Away

We all try and we all fail, at least to some degree. My father has tried harder than most. My father and I continue our conversation about fear, the shameful undercurrent of so much our lives, both his and mine, but about which we never spoke, until now.

"Do you like taking risks?" I ask him. It is still the same day he told me about the Nazis. I know the answer to this—I've been the beneficiary as well as the victim of my father's love of risk. But I'm not sure he knows that he is drawn to danger.

He thinks for a moment. I wonder if he considers risk-seeking behavior irresponsible.

"Calculated risks, yes," he says. "Professionally, yes," he adds, electing to stay on safe ground. "My program was to be canceled at Lincoln Lab [MIT's defense lab, where my father worked for nearly forty years], and I had to pick a new target for the group. I could have followed everyone in the field.

"I would have been following other leaders," he clarifies. "Instead I picked a very difficult target, and I offered that. My sponsors were very surprised. I was proposing to achieve a thirtyfold increase in the capacity of our technology. And then I worked very hard with my associates, and from that point on we could do no wrong. That was a risk, but it was calculated. . . . I believed in the physics."

"Do you like to take risks in other areas of your life?" I ask, naughtily.

"I get curious about how one does things," he says. "Like going into the mountains in wintertime, because you keep hearing stories of occasionally people dying. And I was curious. I got out a handbook, and I went with a group from the Appalachian Mountain Club. I followed along with them." Eventually he would become a certified guide.

My father took us into the mountains again and again. He told us that God is in the mountains, not in the synagogue, and I thought—and still think—he might be right. At least he might be right about one path to the divine.

The night before we left for our mountain trips, we would make a pilgrimage to Eastern Mountain Sports, where we bought supplies and rented equipment we might need in case of ice or snow—gaiters, snowshoes, and crampons. Back then, such equipment was a novelty in America. My father could be cheap, but not when it came to equipment that would keep us safe on the mountain. We would be prepared, or so we thought, for any contingency, loading our backpacks into the car the night before in order to be ready to leave before dawn.

During these trips, it was as if we had pressed a refresh button on our father. He was excited and exciting; pleased with his own self-discipline and strength, and pleased, equally, with ours.

Here is what we learned from our father during that period: if you push yourself hard but pace yourself, stopping frequently to "refuel," you can accomplish nearly whatever you set out to do—a lesson that can take you far in life. My closest childhood friend, who often accompanied us on these trips, became an Olympic rower, eventually winning a bronze medal. She attributed her success as an athlete to the persistence and self-discipline she learned from my father.

I have been wondering whether my father might have pushed us a bit too hard at times. Nonetheless, I am shocked to hear him admit that he was drawn to climbing in winter because he had heard that people died up there.

"When you took us up Mount Washington that time in January, was that the first time you tried climbing it in the winter?" I ask, referring to a trip we took in 1972. I was fourteen years old.

Mount Washington is dangerous to climb in the winter because of the cold and wind, and because of the risk of avalanche. Most hikers stick to trails known to be safe, but we did not. We slid down the snowfields, using our ice axes as rudders.

Mount Washington Observatory warns climbers, "Mount Washington's weather has been called 'the world's worst,' since the fiercest winter conditions on the mountain rival those of the polar regions and the highest mountains on earth." The meteorological conditions are sometimes fatally severe.

It was indeed very cold. Icicles formed on our noses, brows, and lashes.

"Very strong winds are quite common in winter on the mountain. Mid-winter sees daily average winds of between 45 and 50 miles per hour. Typically, two days out of three will yield hurricane force gusts (73 miles per hour or greater)." The observatory lists for potential hikers the kinds of emergency equipment they will need, warning those who are not expert climbers to be sure to climb with experienced guides.

"I'd been winter climbing for some time," he says, "with a group from Lincoln Lab."

I ask him about the way we descended part of the mountain, sliding down the snowfields. "I read later that you are not supposed to slide down the mountainside like that—that it can create an avalanche," I point out, trudging into unfamiliar territory, tacitly accusing my father of something. But what? Frightening us?

The observatory admonishes climbers, "Any winter climber should also strive to learn more about avalanche hazard, how to recognize it, how to avoid it, and should also learn basic avalanche rescue procedures. Avalanche is not an idle or theoretical threat."

"We never went to a place that had avalanches," my father says.

His statement is patently false, but I don't correct him.

"I was taught how to slide down the mountainside—it's got a technical name," he says.

Soon after we commenced our climb, we had to cross a frozen stream. The water was wild, and the ice was thin in places. I lost my balance, and my foot fell through the ice. But we kept going. My father likes to meet his goals. We had the right equipment to keep us warm and safe, but not necessarily the right frame of mind. There were some young engineers from Lincoln Lab who came with us, and perhaps my father didn't want to disappoint them by having to turn back because of a girl's clumsy slip. Of course, I would not have dreamed of requesting to turn back: I wanted to finish what we set out to accomplish almost as much as my father did. Not surprisingly, one of the toes on my right foot turned numb and dark from the cold, but I didn't notice until we got home.

My sister and I fell behind for a time. We ran into each other, each of us walking alone, both of us crying, both of us ashamed to admit to the other that we thought we might die from exhaustion and cold. We were accustomed to our father pushing us to the edge of our physical strength, but this was beyond the edge. Conceding our exhaustion to each other was a shameful admission of failure of will. We trudged on, and soon enough reached the unheated lean-to where we would spend the night. I remember that my father drank cognac from a flask, laughing with his climbing pals. But I was not having fun. I was freezing. I remember wondering how I would get through the night. This was not the most dangerous thing I've ever done, but it was one of the most unpleasant nights I've spent.

The next day, although the temperature was well below zero Fahrenheit, the sun shone brilliantly. If I want to recover a feeling of euphoria, I return in my mind's eye to the sun on that snow, the silence, the sensation of strength in my muscles.

There were many mountain trips, most of them wonderfully fun. We got so strong that we would run up and down the mountain, despite being burdened by old-fashioned heavy equipment and packs. But the winter trip to Mount Washington brings to mind another trip we took, another trip that frightened us.

"How about the time we camped above the tree line in early spring and got caught in a thunderstorm?" I ask him, almost showing my anger now, anger that until now I wasn't aware I felt.

We had camped just above one of the huts that function as rustic inns. When we walked by the hut, which was not yet open, we noticed that workers were in there, getting the place ready for spring. We thought of those huts as serving the wealthy and the weak, people who lacked discipline, people whom—in solidarity with our father—we held in utter disdain. Nonetheless, when it began to storm, my sister and I begged our father to walk down to the hut, where we thought we would be safe.

"When it became clear that the lightning might strike us, I

took the tent down. It had metal poles that could attract lightning," he explains. Somehow, in this exchange, he has reverted to mountain guide and I to a foolish girl, as if my sister and I had been the ones who resisted taking down our tent rather than the other way around.

"The boys in the hut were demanding that we walk down," I remind my father. "They were furious at you for disobeying the rules. They said we were foolhardy, that we were risking our lives."

"It wasn't against the rules," my father insists. "It became illegal in the 1980s. That was bullshit. It was not illegal then," he says, truly angry now at the recollection. "I remember those hut boys giving me all kinds of lip. I knew they had to provide us shelter. That is the AMC [Appalachian Mountain Club] rule," he admonishes me, in lieu of those insolent boys, who are not physically present in the room with us now.

"Did you report them?" I ask, wondering how far my father would have been willing to take his rage back then, even though we truly had been foolhardy, as the "hut boys," as my father referred to them, had said.

"I thought it was trivial," my father says now, swatting at the recollection of the hut boys' "lip" and his pesky, candy-assed daughter.

Candy-assed is one of my father's favorite derisive terms.

"Your memory and mine are disparate," my father concludes, dismissively. "My memory of it was a calculated risk. I didn't think there would be thunderstorms. It was in May. There was no snow and ice at the time."

But I want to return to my father's childhood experience of fear. I have more important fish to fry here.

"Tell me more about what happened when the Nazis went after your mother with the guns," I ask. "Was that time the Nazis came into your house the most frightening incident of your childhood?"

Now that we've dealt with cold and ice, perhaps the Nazis' visitation will be easier.

"I don't know," he says.

He thinks for a moment.

I cannot look at him. I am nauseated again, with the sensation of floating.

"My father was virtually absent in my life from the time I was six until we emigrated," he says.

I had never heard this before.

And now, as I read this over, it occurs to me that my father was virtually absent from our lives after our mother died. We lived with our grandparents until my father married Lisa, and I did not even know, until recently, that my father was living with our grandparents, too. He came in late and left early, sometimes staying out all night, somewhere closer to his office.

"Did you feel abandoned when your father kept disappearing like that?" I ask.

"I didn't have any feeling about my father one way or another," he says matter-of-factly.

He turns to a discussion of the relatives left behind, the ones the Nazis killed.

"I had a strong attachment to my cousin Inge. We did everything together. We made up our own language. She was energetic and full of fun and playful. She was the apple of my maternal grandfather's eye. He favored her very strongly and ignored me completely," he explains, as if that were the rightful order of things.

"Were you jealous?" I ask.

"Yes. She was a free spirit. Spritelike in my memory. I loved her. She was my only friend," he says.

"She and my aunt Ella were sent to a work camp. She would have been eleven by then. And they labored in the work camp. 'Arbeit Macht Frei.' It said that over the gates. Nazis kept very good records. My aunt died in the work camp. But Inge was forced to march to another camp where they exterminated Jews. She was killed in the very last hours of the war," my father says matter-of-factly.

"Normally you would expect your father to protect you. Didn't you feel badly that your father was in hiding, that he wasn't at home to take care of you?" I ask, returning to our earlier topic.

"I heard that he was strong and protective and stalwart. I never experienced this myself—except before the Nazis. I have a memory of him as a strong and safe person around the household. He loved me during that stage, before the Nazis came. He had me help him in the fields. I would lead horses for him while he plowed. He would take me with him on a sleigh ride. Once, when I was around five years old, when we were at the blacksmith's after the horse was shod, he threw me up on top of the horse—it was a huge workhorse—and said to me, 'Ride home.' "

"Were you afraid?" I ask.

"I had all these examples from my brothers."

I have heard this story before, and my father knows it. It is a story with a happy ending, a story about the good effects of pushing children to overcome their fear. Is my father stalling? I wait to see if he will take up my question. Maybe my father reads my silence.

"I didn't feel I was protected," he concedes. "But I don't think I would have felt protected if my father were home. I felt exposed and afraid, and there was no one who could protect me. But he couldn't have protected me anyway. He was under threat. In retrospect, had my father remained there, he would have been killed.

"For four years I was terrorized," my father summarizes.

"Did you ever have another period of terror?" I ask.

He doesn't answer.

"Were you afraid when my mother died?" I ask, pushing myself to broach a topic even more off-limits than the Nazis.

"No. My memory of that period was, she would get fearful and I would try to comfort her and reassure her. I was spending a lot of my time imagining how she was feeling. I accepted cancer as a way of the world, a fact that you dealt with as best you could. But it was a heavy weight on my being, it was always in the back of my mind."

"But you weren't afraid?" I press.

"No," he says. "There were no Nazis waiting in the wings."

"Were you afraid of raising two girls on your own?"

"I didn't quite understand what that meant. I thought about the mechanics of it. After your mother died I was really badly depressed. I couldn't sleep. And I think that persisted for months and months. And I would just work my way though my days. . . . I coped. Not well," he concedes.

My father talks a while about his difficulties after my mother's death. Then he turns to a discussion of my sister's and my characters.

"You were much more willful and determined than Sara. Sara would go with the flow. And you would fight. You were very strong-willed," he says.

"How old was I when you noticed I was strong-willed?"

"You were a lot like Evan," he says, referring to my son. "You were two. No, maybe eighteen months. Maybe even one year old. You loved to turn on the flame of the gas stove. I would say, 'Bad girl!' Eventually I began slapping you. One time you did this, and before I could grab you, you slapped your own hand, called yourself a bad girl, and then blithely turned on the stove.

"That was basically you." He smiles, half annoyed, half bemused. "If you made up your mind to do something, by gosh you were going to do it."

He stops smiling. "After your mother died, you looked hurt and inward. There wasn't any spontaneity to you, and that worried me. You didn't have the quality that you expect to see in children's faces. You went away to nursery school, and you had the expression on your face. And it was very distressing to see. Sort of withdrawn."

"Grandma said I stopped talking. Is that true?" I ask. I didn't believe my grandmother when she told me this, but I never had had the courage to ask my father.

"No. I think you responded when I would talk to you. No spontaneity. You might not have initiated it. And I couldn't stand to see the way you looked. . . . It was painful to see."

At that time, my sister, in contrast, would ask any woman she saw on the street if she would be willing to be our mother.

"Did you think I was manipulating you?" I ask.

"No, not at all. I thought you were suffering, and I had to fix it."

"How did you try to fix me?" I ask.

He doesn't answer. After a while I realize it's time we move on. I need to return to the rape, the topic I meant to discuss with him.

I take a breath and ask him, "How did you feel, reading the material about my rape?"

"I read it. I wanted to read it," he says.

He wanted to read it. I have a dizzying sensation of something between us thawing. Is my father getting a bit soft in his old age? Will this kindness make me weak? But the thaw doesn't last long.

"I was thinking, Here goes Jessie, preoccupied with something that happened thirty years ago, when she could be spending her time and energy enjoying her life," he says, reverting to his more typical mode. But there is kindness to my father's blunt approach, perhaps a truer kindness. "And you do have a prospect of a new life. And I would like you to make a decision to embrace the future and love, and to realize that when you love, you expose yourself to hurt. That is what I thought. I wish I could somehow help you overcome this preoccupation. I felt that you're losing—you're losing life. . . . That is what was uppermost in my mind."

Even now, in the midst of this conversation, my father is trying to fix me.

"For what it's worth, I recognized this detached feeling that you describe," he says.

"And I thought—she can't enjoy her life, " he says, reverting to his previous topic.

My father is switching back and forth, from the engineer who would fix me if only he could, to a person who feels with me.

"And then, once you started reading . . . it was a horrible experience." He is back with me.

"Did you know any of what happened when we were raped?"

"I had none of the details."

"Did you feel any emotion, other than recognizing that it was a horrible situation?"

"You want me to elicit an emotion?" he asks. "I don't have any emotion. I was horrified."

"Did you have to get up while you were reading it, pace around?"

"I didn't want to read it straight through. But I forced myself to read it because I feel I should be witness to horrible acts, no matter how terrible they are," he says.

"Why?" I ask.

"I've always felt that way. To fully understand what actually happened . . . and to bear witness against it."

"And yet you didn't want to do that at the time we were raped?" I ask.

"I couldn't expunge that from your experience. It was there. And you had to deal with it. And my way of dealing with it was to move on," he says.

"What did you feel when you first heard your two daughters had been raped at gunpoint?"

"I didn't dwell on it," he says. "I dwelled on what I could do to make you all right."

"What did you do to make us all right?"

"Talk to you," he says.

I am surprised, because I don't recall my father making any special effort to talk to us.

"About what?" I ask.

"I tried to get your mind off it. Not dwell on the rape. Dwell on things that have to do with life. That would have been my reaction. To try to put it behind you and then move on. And help you with that."

"So you read the material I gave you. Your first reaction was to judge me for being preoccupied, for contemplating my navel, as you used to say." I say this even though I, too, see the seductive appeal of moving on, not contemplating one's navel.

"Yes, that was one of my favorite expressions," he admits. "I've seen people who were preoccupied with the past . . . ," he says, not finishing the thought, but disapprovingly.

"So you thought Sara and I should forget about what happened and move on? As you did, after you escaped the Nazis, and after our mother died?"

"Yes," he says. "And seizing opportunities and life as they present themselves. That would have been my attitude. It still is.

"I believe the best revenge is to live," he says.

This sounds sensible, even admirable. What I love and admire the most about my father is his capacity for joy. My father is drawn to euphoria. He finds those moments of sun so intense that you barely feel the subzero cold. That may be the most precious gift he has given us.

But then I wonder—revenge against whom? Against God for allowing Nazis and rapists?

We move on.

"What is courage?" I ask.

"Bearing witness. That is a form of courage. Accepting reality as best as one knows how."

"When did you come to that position?" I ask.

"All my life," he says.

"What have you borne witness to in your life?" I ask.

"Watching your mother die . . . not withdrawing like some people do. Watching my sister die. I'm going to visit her tomorrow. It is very painful to see her.

"Talking to you now," he adds.

I smile.

"I agree with that," I tell him, firmly. "You are very courageous to talk to me now."

"But I want to help you," he says, apparently not realizing that his bearing witness, for the first time, is help enough.

Still, I feel abandoned by this word *help*. He is pushing me away again. I don't want him to help me. I want him to know me.

"What do you think I do for a living?" I ask, a bit superciliously.

"You teach and write," he answers, puzzled.

"What do I write about?"

"Terrorists," he says.

"What do you think I'm doing in my writing? Would you say that I am bearing witness?"

"Yes, I suppose you are," he concedes.

"Is it truly courage if you bear witness, but you don't feel anything?" I ask.

"Oh, sure, you're right," my father concedes again.

I wonder if he truly believes that.

"What are your most important values?" I ask.

"Honesty . . . integrity . . . the sense that if I make a commitment, I keep it. Steadfastness and the capacity to love. And to enjoy," he adds, as if admonishing me for what he perceives, at least in this moment, as my willful refusal to enjoy life.

"You don't seem to have had an emotional reaction to the material I gave you," I shoot back.

"I wasn't suffused with an emotion," he says. "I thought you were revisiting something that happened long ago . . . it was a preoccupation."

Oh, my God. Again I have to hear this. We are stuck, my father and I. I slash again and again at our mutual denial about what happened and what we would have felt had we been able to feel at the time these events occurred. And he keeps reverting to the accusation that I am preoccupied by the past. I have a notion that if we can admit what happened to us, if we can feel it, we will be freer of demons. But maybe not.

We take a short break. I make us some more tea, and we get back to work. "Why didn't you come home after we were raped?" I ask, finally getting to the question I most want to ask him. My annoyance with both of us has eroded my fear of broaching this topic.

"I think you're confused about the timing," he says. He doesn't seem to remember now what happened. Although he dutifully read the material I wrote about my rape before coming to see me, he seems to have forgotten to read the letter he wrote to me in 1994, in which he explained that he was scheduled to be home in three days. Still, he has excuses for what he now says he doesn't remember.

"Sidney [our physician] saw you at the Emerson hospital," my father says.

"No, he didn't. His partner did," I correct him.

"And he reported that you were well cared for," my father says. "You had Lisa there, and you didn't really want me."

My father had "moved on," I understand now. He was living in his future with his new wife, not in his painful past, when he left my sister and me with his ex-wife, who had not adopted us.

I have always felt that my longing for a mother was childish. Forbidden. It might slow us down. A shameful perversion, right up there with S and M. Probably worse.

Yes, Lisa was caring for us in my father's absence. She was taking good care of us. But our father knew that our relationship with Lisa was complex. We had come to love her as our mother, but she did not always think of us as her children. And who could blame her?

Soon after my father married Lisa, she and our father urged us to refer to her as our mother. She must have wanted daughters as much as we wanted a mother. At first it felt false to call Lisa my mother. I started with "Marmy," from *Little Women*, graduating gradually to "Mom." By the time I was able to utter the word "Mom," without feeling like a Sarah Bernhardt, as we said back then, I had imprinted on her like a duckling. I hadn't noticed that she wasn't in fact a mother duck, through no fault of her own, but a swan.

I was very angry at my father when he insisted, two years before my sister and I were raped, that we move in with him and his third wife, yanking us from the home of our half sisters and Lisa. But it turned out I was mad at my father for the wrong reasons. It was not my father who had initiated this change, but Lisa. My father didn't or couldn't tell us the truth; he wanted to protect us. Our love for Lisa was only intermittently requited, which made us feel the loss of her even more keenly. She had not adopted us, but we had adopted her utterly.

As a small child, I was unusually quiet, a very good girl. But something happened to me by the time I reached puberty. I took on the mantle of difficult child: I was defiant, annoyingly loyal to ideas and persons, skeptical, and tough. I had a hard time accepting Lisa as my second mother, and an even harder time leaving her. But Sara was different: she drank love in, whatever its source. She was beguilingly cheerful and sweet. Anyone would want to take care of such a child, I would have thought. But I would be wrong. I am certain that Lisa loved us, just as we loved her. But we were second class in both households, and would remain so. My father, however, preferred not to notice.

Denial was seductive, not just for us. Despite my efforts to accept the adults' words, the evidence of my senses kept leaking in, and the gap between what was acceptable to think and what a bad, ungrateful girl could not help but observe grew dizzyingly wide. We tried not to notice the obvious disparity in presents for the "real" children and presents for us. But children do notice such things. When I did notice, I felt greedy and ashamed, hoping that my feelings did not show on my face. When Sara and I were not included in a family photograph at our aunt Judy's wedding—Judy, who had been my best friend—I found myself dizzy with confusion. I felt suddenly light and unburdened from the loss of my family, but also alone, unprotected, in a brave new world. Family photographs on the fridges of both houses included our stepmothers' "real" children, but not my sister and me. Of the eight children in the two households, we were the only two who needed braces who didn't get them, but nobody seemed to notice. We were the only ones without a mother to insist.

I tried not to notice. I felt ashamed of feeling pained about such trivial matters—presents and photographs and American-looking teeth. Eventually I came to accept the only explanation that made sense: the grown-ups were good and wise, as grown-ups should be, and I was bad. That conclusion made sense, in part because I was in fact a terror. I would come to see myself as second-class, not just in the context of my family but in the world at large. When good things happen to me in life—and oh, my Lord, they do, again and again—I'm sure there must be a case of mistaken identity.

Even today, typing these words, I feel like a brave but naughty child removing a Band-Aid. Look where I tore off the bandage! There is nothing here but a hypertrophic scar where a dead mother should be. And yet, had I the audacity to mention the word *motherless* when I was a child, it would have been interpreted as a sign of immoral ingratitude and severe instability. In denying that we were motherless, the adults, I am certain, had our best interests in mind. No one wanted to harm us. On the contrary, they wanted to protect us.

"You had projected onto her motherhood. You were going through that myth that Lisa was your mother," he says, disdainfully.

I am astonished that my father would now see my accepting Lisa as my mother, at least as long as she chose to play that role, as a form of what he now calls "projection." I was only doing what I was told. How was I to know that eventually my father would see this mother as a myth?

"But didn't you play a role in creating that myth?" I ask.

"How?"

We have sunk into a kind of cleansing rage, my father and I, divulging secret, shameful thoughts that we would not share in our normal states of mind.

"By banishing the memory or thought of our mother from our lives. No pictures anywhere in the house. I didn't even know that you had any pictures of her until two years ago," I say, in what I now imagine was a childish, rageful tone of voice.

"You never asked," he says, softly.

"But you helped create that myth," I insist.

"In suppressing Shola. Yes, I guess I did."

Silence.

"Why did you do that? Was it just 'Move on—don't think about the past,' just move on to the next phase? Are you sure you didn't banish the memory of our mother because it was too painful for you to bear?

"Did you do this for yourself or for us?" I repeat.

In reading over these notes now, I can hardly believe that I had the courage to utter those words.

"I didn't give it much thought. Decorating houses with pictures of your dead wife didn't seem like a good idea when you have a new wife," he says, apparently not wishing to ask himself how much of what he did "for the good of his children" was actually to protect himself from pain he could not bear to feel.

We move on.

"So are you satisfied that you behaved impeccably after our rape?" I ask, wanting now to end this conversation.

"I behaved reasonably," he says.

"Were you courageous?"

He doesn't answer.

"We were so afraid of you," I add. A non sequitur.

"I was a pussycat compared to what I had growing up," he says. "I guess I would have been fearful of one thing. Of losing you. Losing one's child is a terrible thing."

You haven't lost me, I think to myself. But I need you to feel what happened to you, so that I can let myself feel what happened to me.

I, too, am fearful of loss. Losing a parent, especially this parent, would be crushing, will be crushing. But at least he now knows what happened to me, and I know what happened to him.

No one can take that away. If we lose each other again, to denial or death, at least we will have this moment of knowing.

So my father didn't return home to us right away. Yes, he had work to do; and yes, he was scheduled to return to America several days later, in any case. But I sensed that there had to be a reason that he was capable of compartmentalizing his life in this way. This sort of mystery evokes irresistible curiosity in me. I am compelled to research. But the answers I get are not necessarily related to the questions I pose, and I am not always ready to hear them.

chapter seven

Denial

All the people who knew Brian Beat claimed to be certain he was innocent.

Denial helps the bystander. We don't want to know what the boys we send to Iraq have done to others out of terror, or what others have done to them. We would rather not know about terror or be confronted with evil. This is as true about Abu Ghraib as it is about personal assaults and more private crimes, the crimes that occur inside families.

But the victim, too, cannot bear to believe. She may bury or dissociate from or disown her pain. She may drink or take drugs or become unwittingly promiscuous, compelled to repeat the violation again and again, sometimes in the role of victim, sometimes in the role of perpetrator. The impact of the violation drips lazily down, like that clock in Dalí's painting, pooling in the form of shame. She may remember the facts that transpired, but the outline is blurry. There is a haze in the brain, and the facts are detached from feeling. Certain sounds or scents may terrify the victim, but she may not notice her fear. For me, it's that ticking sound. So irritating. I want to punch. Certain scents, too. But for a very long time, I'd forgotten or dissociated or denied the source of my terrors.

To be raped or abused or threatened with violent death; to be treated as an object in a perpetrator's dream, rather than the subject of your own—these are bad enough. But when observers become complicit in the victim's desire to forget, they become perpetrators, too. This is why traumatized groups sometimes fare better than traumatized individuals. When the feeling of terror is shared, victims have a harder time forgetting what occurred or denying their terror. In the camps, what mattered most, Anna Ornstein explains, was whether there were witnesses willing to share the burden of overwhelming emotion. Talking about what occurred with other survivors or witnesses was an essential part of recovery, Ornstein claims.

When authorities disbelieve the victim, when bystanders refute what they cannot bear to know, they rob the victim of normal existence on the earth. Bystander and victim collude in denial or forgetting, and in so doing, repeat the abuse. Life for the victim now begins anew. In this new world, the victim can no longer trust the evidence of her senses. Something seems to have happened, but what? The ground disappears. This is the alchemy of denial: terror, rage, and pain are replaced with free-floating shame. The victim will begin to wonder: What did I do? She will begin to believe: I must have done something bad. But the sensation of shame is shameful itself, so we dissociate that, too. In the end, a victim who has suffered the denial of others will come to see herself as a liar.

The terrible truth is that once a person has been raped or abused, she seems to acquire a scent or a frequency that makes her an irresistible target for abusers. She may be haunted by a feeling of ungroundedness and by periods of hypervigilance. If she is lucky, as I was, she may find or fall into a career where hypervigilance is useful (though it is unlikely to be useful in her personal life). And if she is terribly unlucky—if she ends up a jailer in Abu Ghraib, for example—she may slip over the edge and victimize others.

The dizziness brought on by the denial of others is often worse than the original crime. When I think about what denial does, I can understand why some victims, thank God a small number, take out a gun and find someone to shoot or maul or rape, sometimes in their own homes.

No one wanted to believe that Brian Beat was a serial rapist of children. Even his jailers. They were in denial, too.

The Massachusetts Department of Corrections gave me a redacted copy of Brian Beat's prison file. Thousands of pages. I know there is a lot of information in these. I mean to read them, but they don't get read. Months go by. A year goes by. Well, if I'm honest, years.

I keep these files in a trash can. It is a small wastebasket, the kind you might use in your living room. I knew that the Concord police had offered me a treasure trove. But I put the files away; I didn't have time to read them. I put the trash can in a rarely used fireplace. And then I forgot about the files.

I look at my trash can, but that seems to be all I can manage.

Once, when I was decorating a house, I went to a sale at the Boston Design Center, the sort of sale where you get useless "decorator" things for 75 percent off. I bought a small wastebasket—a work of art—with three identical elephants enameled on its sides. They have deliciously wrinkled skin, those elephants, and friendly, floppy ears. You can almost feel the impact of their slow, serious tread, which they repeat again and again, around and around. Will they ever get it right? Will they ever be released from this endless circle? The garbage can is too beautiful to use for garbage. I found the perfect use for it: to store the records of Brian Beat's rapes and of the period he spent behind bars.

Eventually, I take the trash can away from where I live and bring it with me to Tanglewood. I open the files in a place I feel safe, sitting outside in the open air, listening to the Boston Symphony Orchestra rehearse. On the day I finally find the courage to begin reading, Thomas Hampson is rehearsing Mahler's *Kindertotenlieder*, a song cycle for voice and a small orchestra. The songs are about a father's coming to terms with the loss of his two children. In the year after his two children died, the poet Friedrich Rückert wrote 425 poems about their death. The lyrics are based on these poems.

This father is writing as a way to process some evil. I think I can understand this. But the father, at first, is in a state of denial. "I often think they have just gone out! Soon they will get back home," he would like to believe, and half does. There is an implicit meditation on the problem of evil: If God is omnipotent, how could He allow these two children to die?

Two children. Our rapist, too, often raped children in pairs. If God is omnipotent, how could he allow these children to be raped? But we didn't die. Outwardly, we remained alive.

The struggle in the *Kindertotenlieder* is with natural evil—the evil caused by chance or by an act of God, such as bad weather or disease; whereas my struggle is with moral evil—the evil caused by men, acting with malevolent intent. But I begin to wonder—can we really distinguish these two forms of evil so cleanly—the moral evil of malicious intent with the natural evil of a storm or disease? Was Brian Beat partly the victim of a storm or a disease?

I did not seek to hear this music—but the musical depiction

of denial, followed by the struggle to sustain hope despite the recognition that evil exists, helped me begin to read.

Brian Beat was convicted of three of the forty-four rapes that the police now believe he was guilty of, and was sent to prison for eighteen years.

I read the statements of a girl who identified Beat as her rapist. (I am not allowed to quote from the redacted statement.) It is June 1972. There were five girls in the house, four of them in a second-floor bedroom. Suddenly a man appeared in the bedroom. The girls had not heard his footsteps on the stairs. They described him as five-ten, about 155 pounds, very slender.

He came into the room holding a gun in his left hand. He told them to be quiet, and insisted that they not look at him. He reached up and pulled the light cord to shut the light off. Then he put a mask on. He pointed to one of the girls and commanded that she follow him. He warned the other girls that if they spoke or did anything, he would kill the girl he was pulling into the next room, gun drawn. He commanded her to take off her pants and to lie down. Then he spread some jelly on her vagina. He told her he was putting the gun near the pillow. After he raped her, he got up and apologized. He told her his gun was really a cap pistol. She described him as gentle, and said she felt sorry for him, even though she had been afraid he would kill her.

These were early days in the rapist's career, two years before he found my sister and me. He hadn't perfected his technique. He did not forbid his victim to speak. He applied that lubricant to his victim's vagina, a thought that nauseates me. But there are similarities. The small pistol with the white handle. His informing the victim after the rape that the gun was just a cap gun. His apologizing. The way he evoked sympathy: I felt sorry for him, too.

The music periodically penetrates my terror with sound. Eventually the father accepts that his children are dead. There is evil in this world, but there is still a God. This father has found a way to recover his faith. Will I, too, recover my faith, even after reading the words of these violated children? The song cycle ends in D major, in a mood of acceptance and transcendence. Will I, like Rückert and Mahler, experience moments of transcendence? Maybe, but not yet.

Now I have brought the files home, back to where I live. I race through them, as if trying to avoid being contaminated. Something has been unblocked. A geyser of fear. As I sit down to read the rest of the files, I am overcome with an embarrassing feeling of terror.

I need to buy some things in preparation for a trip. Sunblock and insect repellent. But I'm afraid to go out on the street. I step out the front door. There is a hum of excitement. Why are these people so unafraid? I feel the warm air on my naked arms. I walk into Harvard Square. It is summer, and the streets are filled with happy people. They are celebrating something. They are celebrating the warmth of the night. I do not like this happy night buzz; it terrifies me. I have the distinct impression that certain people—predators—can read the vulnerability on my face. If I were a young woman, I would be in danger now. Predators would smell my fear.

I return to my apartment. I am not going to fall apart. It is warm, I know that, but I put on more clothing to cover all my flesh. I sit down again to read.

Once Beat was admitted to Bridgewater State, he was evaluated to determine whether he was a sexually dangerous person. I find a letter with the results of this evaluation.

Dear Sir:

At your request, I examined Brian X. Beat, N-21668, at the Mass Correctional Inst at Norfolk, on April 12, 1974. The purpose of the examination was for psychiatric evaluation to see if he may be a sexually dangerous person under Section 6, Chapter 123-A, G.L.

The inmate was informed of the nature and purpose of the examination and was also informed that the examination was not privileged. The inmate is a 27-year-old single man, sentenced on October 10th, 1973, by the Barnstable Superior Court, to 12–20 years for armed assault and 12–20 years concurrent for Rape.

The police version of the present offense as given in the record states that four women were sitting in a 2-bedroom apartment at 1:30 AM on June 20, 1972. A man appeared wielding a small pistol. He immediately put out the light and proceeded to put on a mask after asking questions to four of the girls in the one bedroom. He pointed a gun at [blacked out], told her to go upstairs with him, told the others if they called for help someone would get hurt. He then had sex relations with [blacked out] under threat of harm. She stated that he was very gentle and hurt her in no way. They then returned downstairs and he left. He was then identified by the witnesses as the person who had committed the B & E and the rape.

There is no evidence of mental disease. He was found guilty of this one offense, which he is now serving, but the circumstances do not point to marked aggressive behavior. It is therefore my opinion that he is not a sexually dangerous person. He does have other charges pending on sex offenses and if he should be found guilty on these charges, he should be re-evaluated taking the nature of those offenses into account.

Sincerely yours,

Carl Henks, M.D.

I read this letter, and I have to get up. I wash the dishes. I had soaked some dried lima beans overnight, and their skins have become loose and wrinkled, like tiny wrinkled foreskins. It seems to me these skins need to be removed. I gently peel the skin from every bean—a job I would normally find painstaking and annoying. I am satisfied by the clean white interiors; smooth, vulnerable, harmless.

It occurs to me that I would like to take a baseball bat to this man, this so-called psychiatrist, Dr. Henks.

"He then had sex relations with [blacked out] under threat of harm. She stated that he was very gentle and hurt her in no way."

"Sex relations," he writes! As if it were an unremarkable occurrence that a masked man with a gun would have "sex relations" with a girl.

Can a trained psychiatrist really assume that a rapist with a gun in his hand could have "sex relations" with a girl under threat of harm and still "hurt her in no way"? How can he repeat the petrified girl's use of the word *gentle*? What does the doctor mean when he says the rapist "hurt" her in no way? Does he conclude that, because Beat did not tear his victim's eyes out, did not bloody her limbs or break her bones, that he did not hurt her? As I write these words I imagine this doctor's penis wilting and shrinking in terror, as small as a bean, and there is some satisfaction in this cruel thought. But wilting is not enough: I want to bloody him. In my mind's eye I swing a bat right at this doctor's learned head, smashing his skull, the skull that contained his bad, addled brain, a brain capable of judging a convicted rapist as a not sexually dangerous person. I also take the bat to the part of him that had sex relations. I am a good girl, of course, so the doctor would not expect me to harm him.

But now that I've written these words, I retreat in horror at my own violence. I like to think of myself as civilized. I am seeking transcendence, not violence.

But there is a perpetrator inside me. I feel a kind of adrenaline at the thought of harming this man. Even as I'm ashamed of this ugly side of myself, I remain angry enough with this so-called doctor that I will leave these words here on this page. I will display my shameful thoughts to ward off future rapists and their protectors, to make clear how women may ultimately see them, to make these men cross their legs with a premonition of fear.

I read the letter again, finding myself a bit confused. On the one hand, the doctor appears to believe that a rape occurred. "He was found guilty of this one offense, which he is now serving." The doctor does not appear to question the guilty verdict. He just doesn't find rape at gunpoint to be an example of "marked aggressive behavior." And he finds it possible to describe a convicted rapist—whose guilt he appears not to question—as "not a sexually dangerous person." What is the matter with this guy? Now I wonder, would he expose his daughter to this "gentle" gun-wielding rapist? If so, would he ensure that his daughter was armed? What about his wife? Now my thoughts run embarrassingly back in the direction of violence. I would like to shoot the two of them, both the learned doctor and the "gentle" rapist displaying "no evidence of mental disease." But the rapist is now dead, and the good doctor is probably dead, too.

What sort of person fantasizes about shooting doctors, or lopping off their private parts?

He is conducting this interview with inmate Beat under strange circumstances. He wants the inmate to talk, but he must report what the inmate says; and the law requires that he inform the inmate of the risks to himself. This is exactly what I have done, countless times, with terrorists. I want them to talk to me about their violent crimes, but I write about them; and ethics requires that I inform them of the risks to themselves.

I return to my lima beans. I pull a barely used cookbook off the shelf, an old Fanny Farmer. I find a recipe for baked lima beans. I boil the beans and drain them. I parboil a head of garlic to make it easy to remove the skin; my taste is a bit more robust than Ms. Farmer's. I put the skinless beans and the skinless garlic in a pan in a hot oven and leave them there to roast while I type up my thoughts. I think of removing the skin from my rapist's penis, to reveal a trembling sea slug; a man petrified of me, petrified of what I might do next. Perhaps this garlic will cure me.

I have just read this passage out loud to my sister. My sister knows that I'm doing this work, of course. But I don't send her too much of it to read, as I'm afraid to retraumatize her. But this rage, that seems okay to share.

"Your imagination is too vivid," she tells me. She fears my readers will conclude that I am mad, harboring such violent fantasies. Here is my answer. Any person who has experienced acts of extreme violence will have such fantasies, though they might forget them. I took sexual violence into my body, and it became a part of me. It is better to know one's shadow side than to pretend it doesn't exist. Fantasizing is very different from acting.

Unbelievably, a second doctor concurred with these findings. Although Brian Beat was convicted of rape, this doctor altered the crime to "attempted rape," confirming that Brian Beat was not a "sexually dangerous person."

chapter eight

The End of Denial

John Henry was living with Brian Beat on the Cape on the day that the police arrested him for rape. At the time, John told the police that Brian Beat was innocent. Now he isn't so sure.

John is the last member of Brian Beat's circle of three closest friends. Abby, Brian Beat's high school girlfriend, told me about John, but she thought he was dead. I don't know why she thought that.

The police told me they thought John was alive. He had been a witness to a murder several years ago, in Worcester. But they didn't know where he was living. Jack, my research assistant, was the one who managed to find him. It turns out that John often lives in the woods in a tent, but he was staying at his sister's house at the time we found him. We started sending John letters, both by mail and by fax, but it took him some time to respond. It seems that he wanted to research me first. When we find him, he will tell me that he took my books on terrorism out of the library, and is now reading the second one.

Chet drives me, as usual; and as usual, by the time we arrive, I am so sleepy I can barely focus. I am tired of this sleepiness, tired of imposing on Chet in this way. He is tired of it, too, he confesses.

John lives in a one-room house at the end of a dead-end road on the edge of the lake. A summer retreat. Out the window I notice that there is very little here now that summer is past. An emptiness. Today it is bleak, with the cold white light of the late fall reflecting off the lake, and a bitter wind.

We pull up next to a pickup in John's driveway. He hears us arrive. Not a lot of visitors at this time of year. We see him come out to the front of the house to greet us, a thin man with weathered skin, in a heavy jacket. He is followed by a half-blind dog, so arthritic she can barely walk. The dog barks hoarsely. She means to protect her master, but she can barely summon the energy. I find myself slightly alarmed by the dog's unfriendly bark, while simultaneously distressed by her weakened state. "Don't worry," he says. "She'll settle down. After she gets used to you."

He offers me a rough, red hand. I am surprised. For some reason I expect him to feel that I might contaminate him. He is not like Abby or Simon. He seems self-contained, even refined. A refined recluse. We are here on a Sunday because John works as a landscape contractor the rest of the week. He urges me to sit next to the wood-burning stove. He's concerned that I might be cold. He offers Chet a chair nearby. The dog whimpers as she settles herself at John's feet. Life is pain, I think to myself.

I ask John how he first met Brian Beat.

"We went to school together at St. Louis. Starting in 1955. I was in the third grade."

He speaks slowly, carefully, trying to get this right.

"I first met him on the playground at St. Louis. I was eight years old. Maybe nine. He was bigger than everyone else. He was choosing people for his team. I remember that he was smiling. He had a nice smile. You want to be on my side, he said. But then he suddenly punched me so hard I lost my breath. . . . He was still smiling. I don't think he meant to hurt me. I was stronger than he was, so it was strange that he did this.

"My sister Cathy knows a lot more about his family. My sister lived with Brian's birth mother from the time she was eleven. His birth mother was my sister's foster mother. You should talk to Cathy," he offers.

I ignore this reference to Cathy. I worry that he will want us to leave before I learn what I need to know. "But you became friends," I ask, "despite his having punched you like that?" I can see that John is a loner. "I wouldn't say I was friends with him back then," he says. "I don't know why he punched me. It is still a mystery to me. Most of the time he was nice. Ninety-nine percent of the people who knew him would probably say he was nice. But I sometimes saw a side of him that wasn't nice at all.

"I was with his pack briefly and then decided I didn't like what was going on. There was an apple tree near our school. He would try to get us to throw apples at the school and at other kids. He'd be smiling all the time, even when he was doing something like that. I got tired of it. Brian was always doing things that would make you mad.

"I spent more time with Brian later. I became homeless when I was sixteen years old. That's when we started hanging around together as a group—Simon Brown, Brian Beat, and me. That's when we started getting more into drugs."

A memory from my own teenage years intrudes into his story. I was sixteen, the year after I was raped. I was gathering my books for school. I was planning to ride my bike. I was late, half rushing and half dawdling, the way kids do. Rushing, but also curious about what the house would feel like empty of kids. No one knew I was home. I could hear my stepmother in the kitchen. I heard the crash of pots being put away with an efficient hand. An angry hand. She was telling my father that she was fed up with me. She had had it up to here. I was fed up with me, too. Too many mothers. I knew I was difficult, but I couldn't seem to help it. I ran away. I spent the night in a graveyard in Concord center. I'm not sure why I wanted to sleep among the dead. This wasn't the only time I ran away, but it was the only time I slept in a graveyard. Those grave markers built into a hill. I can still feel the old gray stone, worn thin by the wind. The smell was bracingly cold: the scent of cold, gray facts. The fact was, she did not like me. The fact was, I was difficult. The fact is, you are always alone when you die. But there was that other scent, too—the scent of grass that tufted behind those thin gray grave markers, some of them from before the Revolutionary War. Somehow, the knowledge that people have been dying and living for hundreds of years was a comfort.

Now, the sound of John's voice brings me back into the room. "My stepfather was an alcoholic, and he would come after us kids when he got mad. My mom and my real dad split up when I was five. Then my mom married my stepdad. He was going to shoot my mom and my two sisters and me. . . . My parents were having money problems and fighting and my stepfather had us all sitting on the couch and said he was going to kill us and he had his shotgun. He was holding the gun. Running around screaming at my sisters and me with his gun in his hands. He said he was going to kill us, but I wasn't going to let him. I was strong. Stronger than him. I grabbed that shotgun out of his hand. I went after him with a baseball bat. I threw the gun into the middle of the millpond. When the police came, they found the gun there, in the pond.

"That's when he kicked me out. He kicked all of us out of the house, including my mom." I catch my breath. I try to push down whatever chemical is rising in my veins, but it is hard. Like trying to will mercury in a thermometer not to go up when the temperature is rising.

John's narrative comes to me as a shock, though I don't know why it should. The children in Brian Beat's circle were all seriously abused in some way. But I have lost my tolerance for their pain.

He continues.

"I'd been watching my stepfather beat my mom since I was eight years old. He would get into these heated arguments with my mom, under the influence of alcohol. If I started telling you the grisly details—"

He pauses.

"It would be the whole book," he says, finishing his sentence, but not his thought. I can see that there is some awful memory plaguing him that he wants to unburden himself of. He wants to tell me, but he is afraid that I won't be interested. I sense this, but I do not help him. I just listen, trying to stop myself from empathizing. Writing this now, I feel ashamed that I did not want to hear any more. I want to tell him that I, too, suffered, even though I was never homeless, and even though I look so well cared for and well fed.

"My sisters and I. We used to huddle together and cry under the covers at night. Whenever it was possible, I protected them. I was the oldest. They black it out. They've forgotten. I'm sort of jealous. . . ."

"I'm an oldest sibling, too," I tell him. I don't voice the thought that I, too, saw and remember more than my sisters do. I, too, saw and remember more than I wish.

"One time they were fighting so severely. There were French doors. I hate French doors to this day. I was watching through the French doors. He stuck her on a hook. I saw my mom on that hook. I was nine. I ran up to where she was. I had to take her off the hook. She fell on me. We were lying there in a pool of fluids and blood that was coming out of her," he says.

A pool of fluids. Terrible thoughts come into my mind. Perhaps her eyeballs were leaking. Perhaps an organ got caught on the hook. Perhaps an eardrum had burst, releasing a thin yellow stream. It was probably just blood. The sight of blood and tears would be shocking to a nine-year-old boy, I tell myself.

"I would never do that to someone," he says. Why does he tell me this? Is there a someone he would have liked to hang? Is he worried I might think he would?

"I wish it never happened," he adds.

I have a sense that there is more to this scene than he remembers, images or sounds or scents that he could not process at the time.

"Of course she stayed with him after that," he says, bitterly. "She had that syndrome."

At this moment, neither of us remembers the phrase *battered-wife syndrome*.

"She said she married him for the sake of us kids," he says, skeptically. "I think it would have been better for all of us if she had remained single." An understatement.

In spite of myself, I am drawn into the story. I am with this nine-year-old boy now, who was so determined to protect his little sisters, those sisters who cannot even remember what happened.

"My sisters have successfully blanked this out," he repeats.

I see that John is alone on this earth, that he is haunted by the image or sound or scent of those fluids, whatever they were, even at age sixty-three.

"I know that those kinds of experiences could lead me to be very violent. That is why I became a vegetarian and started meditating," he says. He tells me that he started meditating as a teenager.

"I don't get close to people. I don't enjoy the give-and-take of close personal relationships. He travels fastest who travels alone." I don't respond, still overwhelmed by his story.

"You know that saying, don't you?" he asks. " 'Down to Gehenna or up to the throne, he travels fastest who travels alone.' "

I tell him that I recognize the words, though I don't know what they are from. He's not sure either. Later I will find that the longer version of the proverb comes from Kipling. Gehenna is a Hebrew word that refers to hell, the place where evil is ultimately destroyed.

"We all got kicked out," he continues. "My mom kept us together for a year and a half. But she wasn't up to raising us on her own, and one day she just disappeared, leaving us behind.

"So it's during that period, when I was homeless, that's the time when I spent a lot of time with Beat. We did a lot of drugs. I regret that now," he says.

"What kind of drugs?" I ask.

"Brian's girlfriend was working in a nursing home, and she could steal prescription drugs. She could steal anything. We always had pills. . . . We could get any pills we wanted, for a long time."

He must be referring to Abby, I realize now. This is what her brother was talking about when he said that Abby had laid her own bed and now she had to lie in it. That terrible bed.

"I was homeless, so I was spending time with these guys. It got me in trouble in '66 or '67. We had a party at a house that I was living at, and the police came looking for drugs. I had to go to jail for a few months for that. Jail was a wake-up call for me.

"After that I moved to the Cape. I got off drugs. I had calmed down. I had a day job, cleaning up a restaurant. I was living in a tent. It turned out that Brian was living down on the Cape, too. One day I ran into him. He asked me where I was living. I told him. He said to me, I got a place, I could use a roommate. He had rented a small house. He seemed to be doing really well.

He had a nice job. He was working long hours as a plumber's apprentice.

"Brian was strange, though," he says, returning to the topic I came here to discuss. "He had a mean streak."

"What do you mean by a mean streak?" I ask.

"He wasn't mean the way I picture rapists."

So he knows that Brian was guilty.

"It was as if he had a dual personality. He was always smiling. But he could be sadistic."

He thinks of an example. "We used to do intravenous drugs—we prepared them out of the drugs that his girlfriend brought us. We made opium out of paregoric. It was used to put on baby's gums when they are teething, but it has opium in it, so if you had enough of it you could extract the opium. We had all done it enough so we knew what we were doing. It was easy to find the vein and stick the needle in. But there was a girl with us who didn't know what to do. Brian offered to help her. He missed the vein. If the drug goes into your muscles, it burns. He burned her arm very badly. Her arm swelled up, but he was smiling the whole time.

"Brian had this grin on his face," he repeats. "We all agreed later that he did it on purpose. He seemed to enjoy it."

He pauses.

"I would say he was disturbed," he says. That word again. The same word Mary, the Beats' neighbor, used to refer to Brian Beat.

Now he has some misgivings. There is a part of him that is like Mary. He, too, would like not to speak ill of the dead.

"I feel disloyal telling you all this. He could understand the beauty of a beautiful poem, but he couldn't arrange his life so he could live that way. He could envision the beauty of life and the goodness of life but somehow wasn't able to actualize what he knew was possible. It's a shame. Like I said, most of the time he was a nice guy. It's just that one of the gears didn't mesh. . . ."

I don't respond. I have little tolerance for this kind of spiritual mumbo-jumbo, especially in this context.

Then I tell him. "I think Brian Beat raped me." I tell him this apologetically, tentatively. I am worried about how he will react. This is an experiment for me, to tell someone who knew Brian in his youth that I knew him, too, but in a very different way.

"I didn't know that!" he says, covering his mouth, as if wishing he could take back everything he said until now. I see that he feels responsible for not knowing, that he might have tried to alter the way he talked about Beat or himself had he known that I, too, was a victim, even though I write books and teach at a famous university.

"How could you possibly know?" I ask, in what I hope is a sympathetic tone. But in truth, I am happy he feels ashamed. This is the first of Brian Beat's friends who hasn't insisted that Beat was innocent, and I am relieved. But not totally relieved. I want him to suffer, at least a little bit, for the denial of his childhood friends and for that spiritual mumbo-jumbo, too.

The room is now permeated with shame—his story of drugs and homelessness, my thoughts about running away, my unbecoming desire to hurt John. Even the dog is ashamed of her paltry bark. And now that I've divulged my rape, there is the presence here, too, of my violated vagina. As I type this word, I cross my legs. I cover my vagina with a thick metal shield. Titanium with spikes. I surround myself with armed guards.

Now that my rape is in the air between us, I want to leave immediately. I want to stand up and walk out that door. I will myself to remain seated. I justify my claim by telling John about the police reopening the case, about the gun, the unusual modus operandi.

"Are you sure you want to talk about this?" he asks.

"I'm sure. Don't worry," I lie, trying to make light of all this.

But there is something new here. I cannot pretend that I'm so strong that the rape means little to me, the way I used to.

"Tell me more," I insist. "More about his mean streak."

"One time on the Cape, I was walking down the street. He was walking toward me, eating an ice cream cone. I noticed he didn't look like himself. When he reached me, he bit me, hard, on the cheek. He hopped on my back and put the ice cream cone on my head. All the while he had a grin on his face. But he looked demonic. I was scared. He was a lot bigger than me. He wasn't thuggish or churlish, but he could be crazy looking. It came on suddenly. Some of his behaviors were crazy."

"Did it hurt when he bit you?" I ask, shocked by the bizarreness of this story.

"A lot."

"What did you say to him?" I ask.

"I was scared. I didn't know what was going on," John tells me. "Sometimes he was strange. Scary. He was thin, but he was always a little bit taller and strong. Not big, but strong. And taller.

"One time we were driving. We had just drank cough syrup to get high. We hit a dog. We yelled at Brian to pull the car over to help the dog. He didn't want to pull over, but we insisted, and he did. The dog was dying. I felt sick. But there was nothing we could do for the dog. Brian didn't seem to care at all. He wasn't upset like the rest of us."

"Tell me about what happened when you were living together on the Cape."

"He was always coming and going. Some of these jobs were late-night commercial work. So he was always coming home late. I never knew where he was. That's why when he tried to get me to be his alibi, I wasn't really sure if he was home when he said he was. He told me that some girls had been raped, that the police might be coming to ask some questions. He told me that I should tell them that I remember him being home, that I was watching the news on the couch in the living room. I think it was the eleven o'clock news. The rape must have occurred during the newscast. It may have been half an hour before or after. He wanted me to say that he was there. Truthfully, I really didn't know. I had dozed off.

"I went to Hyannis, to the courthouse, to testify. He had this really inept lawyer."

"In what way was he inept?" I ask.

"He was really unimpressive. He wanted me to testify, but I sat in the court all day. They never called me."

"How did you come to believe he might be guilty?" I ask.

"Just reflecting on the whole thing. . . . It just felt right that he could be a rapist. I can't say it was any one thing. There was a wildness to Brian. He was not a fighter or a bully in my experience. But he had this wild spurt of sadism. . . . The time he bit me. The time he punched me in the stomach, still smiling. The time he missed the vein on purpose. Always smiling. You add it up. There was a bizarre element to him. And one of the girls who had been raped, afterward I talked to her. I was defending Brian. And she said, 'That man raped me.' Somehow, I understood then that it was true."

He is thinking now. "I wonder if he had any paraphernalia in the house," by which I assume he must mean the mask and the gun that Beat used. He used a wig, too, for the rapes on the Cape.

"I am not a nosy, meddlesome person. I never went in his room. But he had a truck—a panel truck. It was the kind they used to use as a bread truck. It was sinister looking," he recalls. "They didn't have the windows like the vans do now. There might have been a little side window. He might have hidden things in there."

The truck seems to make him think of something else.

"One thing I thought was strange—Brian used to like to go out late at night in these real short shorts. Jeans with the legs cut off to make shorts but cut real short. Like a woman would wear. I thought it was strange. Too short for a straight man to wear in Provincetown."

"Do you think he might have been homosexual?" I ask.

"I wouldn't say he was gay in general. Girls were attracted to him . . . a lot. He had no trouble getting dates, wherever he went. He seemed so smooth with girls. I was kind of jealous. I wasn't smooth at all.

"But he liked to go to this gay bar with his short shorts. He would make an excuse that he was going there to swim. It was a gay bar at a hotel. They had a pool. The pool had lights, and it was open at night. I think it was called the Crown and Anchor. It was the last bar on the left if you were going down the strip toward the water.

"It was fine to be straight in P-town back then because you knew where to go. You wouldn't go to the gay clubs because you knew they were gay clubs. I always felt like he went there in his tight shorts not just because he liked to swim."

"Did you know that Brian got out of the military by saying that he had been homosexual?" I ask.

"No, I wasn't aware of that," he says. He looks surprised.

But then he thinks of something else.

"Years later, after I hadn't seen Brian or Simon for a long time, I started getting more into the folk music scene. I was pretending I knew a lot about it. Brian and Simon used to go to this club in Boston all the time—I think it was called the Loft. They said that they went to hear folk music. I never went with them. But a few years later I was trying to impress some people, trying to make them think I knew about folk music. I told them I used to go to the Loft to hear folk music. They all laughed and said that the Loft wasn't a folk club but a gay bar. I was embarrassed, but I said, No, no, it's a folk bar. It wouldn't surprise me if it was a gay bar and Simon and Brian went there trolling, looking for older men to buy them drugs or whatever." The Loft was in fact a gay bar back then.

"Did you ever hear about abusive priests at your elementary school?" I ask, changing the subject.

"I was naive about that. At that time families went to priests for help. You wouldn't suspect them of anything. But one time my mother and stepfather sat me down and asked me pointedly if this one priest ever did anything to me. There must have been rumors going around about him because my parents asked me in a funny way. Pointedly, like there was a good reason for their question."

"Did you ever hear that any other boys might have been abused?"

"No," he says. "I never heard anyone talk about that." But then he tells me about a boy at St. Louis School who was openly gay. "It was unusual back then," he says. "His family was involved with the church. He had a few brothers. One brother hung himself. In Webster. The brother who hung himself, he was gay, too."

I wonder why the brother hung himself. Could he, too, have been abused by a priest?

John urges me again to talk to his sister. He gives me her number. As we are leaving, he says to me, "I have never told anyone that story about the hook. I'm glad I told you. I feel much better now."

We have exchanged gifts, this recluse and I.

One of the gifts he gave me was the recommendation to call his sister Cathy. When I did, I discovered that Brian Beat had tried to rape her, too. He was in high school at the time. She managed to fight him off. At that time, he hadn't yet thought of threatening his victims with a gun, at least not with her. At last, I have found someone able to confirm that the sadistic rapist that I think I recall was real. At last, I have found someone who doesn't feel the need to protect herself—and punish me—with denial.

chapter nine

The Witnesses

Research is unpredictable. You can chase some subjects seemingly forever and hit a brick wall; but sometimes serendipity strikes. I wanted to interview Hamas, for example, so I went to Amman. I left Amman without a single interview. But when I got to Gaza, two important leaders were willing to talk to me. I've learned to accept, even to expect, that luck will play a big role in my work. Even so, Jack's coming to work for me and what he has been able to help me see have been a surprise.

I was away when Jack began to search for John in earnest. First he drove out to Worcester, John's last known address.

"I went to his mailbox," Jack told me. "The mailbox at the address I found for him. I thought about opening it. But then I realized it might be illegal, so I called my friend Julian. He's a lawyer. Julian told me it's illegal to go through someone's mail. You can look, but you can't touch. So I looked. I saw that John's roommate owned a cleaning company. I ended up looking up the company. I found another lead by looking into another mailbox without touching. It was one of those mailboxes in an apartment building, so the top was stuck up. I went to the other address and there was nothing there. I came back and told you we had some leads. I went to the third address, and there was a car there. It was a truck, a pickup truck. I sort of staked out the scene. I drove by his house a few times. To get the license plate number. I didn't want to stop and be seen. But I'm not sure if it's his house. So I called my friend who works for an insurance company. She has access to the RMV system. I said to her, I need a favor. I need you to look up this license plate. And it was John Henry's. I wondered, should I knock on the door. Then I saw him."

Jack kept searching for John, long past the point where I have given up on finding him. Why was he so determined?

First I better tell you how I found Jack, and how he came to work on this project as my research assistant. A friend of mine, Jen Lockwood, who is an artist, bought a share in a storied Irish pub in Cambridge called the Plough and Stars. The bar was famous for offering Guinness on tap, long before it was available anywhere else in America. It is a rowdy bar that attracts a mix of locals—housepainter poets, construction worker artists, drifter intellectuals, academics from Harvard and MIT, and politicians. Jarrett Barrios, who had once worked as a bartender, launched his campaign for state senate at the bar. It was the birthplace of *Ploughshares*, a leading American literary journal, in 1971, but had recently fallen on hard times. Jen put a lot of energy into reenlivening the bar. She remodeled. She painted the walls a velvety red and hung silvery stars from the ceiling. She brought in her husband as the new chef. And she hired Jack as her bartender.

Jack is particularly well suited for the job at the Plough and Stars. To begin with, he's Irish American. He has "good court vision," one of the bar's owners told me. He might be joking with several customers at once, while simultaneously twinkling a lupine eye toward a regular just pulling up a chair. Jack is a graduate student during the day and a bartender by night. Jen suggested that I hire him as my research assistant. He's smart and a good worker, she told me, and he's involved in terrorism studies. Could I have him e-mail you? she wanted to know. I said yes.

When I told Jen the subject of this book, she told me that she and Jack had just had a conversation about Jack's father, who was in the marines, and had come back from Vietnam with what Jack suspected was post-traumatic stress disorder (PTSD). His father had died at age forty-eight, when Jack was twenty.

"Jen and I had this conversation about my father," Jack told me later. "I don't normally present myself as the son of a PTSD case. My father's been dead for so long. It's not something that I think about that much," he said. "And anyway, it never occurred to me it would be relevant. I thought I was taking a job to help you with your work on terrorism."

Of course I would have to check out why Jack would think his father had PTSD just because he had been in Vietnam.

"What makes you think your dad had PTSD?" I ask him.

"My dad was stationed in Da Nang when he wasn't out on patrol. Da Nang was constantly being attacked. The U.S. forces would shut down their outgoing artillery when the base came under attack, so that the attackers wouldn't find the rocket launchers. When the constant noise of the outgoing artillery stopped, it meant you were under attack. When my father returned from Vietnam, he would sometimes bolt if he heard silence. If he woke up to silence in the middle of the night, he would scatter for his gun. The only way he could sleep is if he put the radio on static. He would tune the radio to static and crank it all the way up. If my mom tried to turn the static down, he would wake up in a panic. My mom tried to wean him off the static sound. She turned the radio down little by little."

Jack's father had PTSD.

His father would fall asleep on the couch on Saturday afternoons, Jack tells me. Sometimes he would wake up confused and in a rage. "We never knew when, but when it happened, it was really alarming," Jack explains. "It was years before we were old enough to understand what was happening with him.

"It's not representative of his behavior—much of the time he was a fun, wonderful father, but you never knew when he'd come into the room and demand ridiculous things of us. He would shout at us. We were afraid of him. Really afraid. He would have random checks. We got hit hard when we had it coming. He threw me through the closet door once. After that I made sure I didn't misbehave again," Jack says.

Jack was reluctant to tell me that his father beat him. But once he got going, it was hard for him to stop.

"He was scary. All my friends were afraid of him. All my cousins. But the upside is that he trusted us. When we lived on Staten Island, we could do whatever we wanted. We were encouraged to go wander around in Manhattan. We did all this crazy stuff. He would say, you better learn how to survive in the Bronx. Take your skateboard over there! We could go over to Alphabet City [a part of the East Village in Manhattan that was dangerous in the 1980s, when Jack was growing up]. I always remember feeling more street-smart than all my friends. All the other kids had to be home by ten. We could stay out as late as we wanted to. But the next morning, we had better be ready and in shape to do whatever he wanted us to do that day. He would come in our room at five thirty in the morning, shout 'Reveille! Reveille! Reveille! Feet on the floor!' Sometimes he would run us. My brother and me. Sometimes he'd have us help him fix our latest broken-down car. 'Shine the goddamn flashlight where I'm working!' he would scream. And we had better jump to do what we were told."

Was Jack's father training him to survive some kind of war? I need to know. It all sounds eerily familiar.

"Run us? What do you mean by 'run us'?" I ask.

"Make us run. He would run right behind us. I was strong and sort of submissive. But my older brother was a little bit less athletic, and he didn't want to run. He would be crying. My father would run right behind my brother. He would be behind him and kick him if he slacked off. We didn't run very far, a three-mile loop. But my dad would try to run us fast. He was a marathoner. He ran the NYC marathon when he was forty-six. He ran a lot of 10-mile road races."

"Then my father had a problem at work. He got demoted. Right after that he got sick. It was clear he was suffering psychologically. He had all this fear. He got cancer. When I'd stay with him in the hospital, he would wake up in the middle of the night and start mumbling incoherently. He would be very afraid, and I wasn't sure why. 'I've told you too much,' he'd repeat several times, and then fall back into his morphine-induced sleep. But he didn't really tell me anything. A month later he was dead."

I see that Jack is reflecting.

"But I think the bigger part of his trauma is that my father was beaten by his father. My father would talk about it sometimes, as if to say, 'You got it easy compared with what I went through with my old man.' My cousins told us that my grandfather beat my uncle so bad he got brain damage. And my grandfather's father beat his kids. The family myth is that our great-grandfather beat his son to death. That's a secret scandal, a family secret. No one was ever prosecuted. That was what you did. You discipline your kids. It's an Irish thing. My uncles that are still alive have started talking to me about it."

An Irish thing? I think to myself. By now, I am surprised that it took so long before Jack came to understand the truth of why he was so driven in this work. He finally got it, he tells me. "It was about my dad. Plus, my girlfriend in college. She was gang-raped," he tells me.

"What do you mean, gang-raped?" I gasp. I am not sure I want to hear this. A sleepiness washes over me. I consider leaving this part of Jack's story out. It's too much for me and too much for the reader. But it's an important part of Jack's life so I've left it here.

I notice a flutter of pain on Jack's right cheek. But he continues his story and I don't stop him.

"We were in college. She was at a party. She drank too much. She was really drunk. Some boys took her into a room and took turns with her. After that we had a really hard time. I was only eighteen. I hadn't had a lot of sexual experience. After that I was scared for her to go to parties. Scared for her to drink. Scared to tell her not to drink because I didn't want to be paternalistic. And so on. Later she told me she'd been abused as a child. Her stepfather used to walk around at night naked with a gun in his hand."

"Where was her mother?" I want to know.

"She says her mother didn't know. Her mother was in denial," he says. I see from the look on his face that Jack is suddenly aware of what he's been telling me.

"I didn't want to tell you any of this," he adds, apologetically. "Sometimes the job became traumatizing for me, but I was afraid to tell you." He sighs.

I feel my shoulders tensing. I was afraid of this. I want to know, but also don't want to know.

"What do you mean? Why didn't you tell me?" I ask.

"I didn't want to express any vulnerability to you. If I told you the work was traumatizing, I knew you wouldn't let me do it. You might push me away. I felt I had to do the work without flinching. I figured I just had to tough it out. Interesting jobs are going to be tough," he concludes.

"I'm so sorry," I say. "You should have told me."

Jack has a girlfriend from Turkey that he's been living with for five years. He is planning to marry her.

"I didn't want to tell my girlfriend about what I've been doing for you," he says. "It seemed to me it was unprofessional to talk about it. And I didn't want to bring it home. She comes from a world where rape is still the victim's fault. It would annoy me if she blamed you. When I finally told her about this project, she reacted strongly. She started crying. It made me worry that she would tell me that she was raped, too. I didn't want to hear that. This is awful, but I didn't want to know. It would just wear me out. I was fully prepared to let her act as though she had gotten over it. Women who say they got raped can get killed in Turkey. Honor killing. Or forced into marriage to their rapist. So no one talks about sexual violence there. She had never heard about rape or sexual violence, except in an abstract way."

A grandiose thought comes to me: This is why I have to write this book, to speak out for those who cannot speak. I push the thought away.

"So I had these experiences that made me understand what you had been through. And also, I knew how to interpret the way you might be acting."

I'm stunned.

"What do you mean, the way I might be acting?" I ask. This is not where I expected this conversation to go.

"I understood when you were in the zone. I had seen it happen to my father a lot."

" 'In the zone?' What do you mean by 'in the zone'?" I ask.

"You just get different. You seem cold, officious. I've seen it happen lots of times," he says.

I feel my toes curl up in my shoes.

"I first noticed when we went to see Stevie at Global Gas in Milbridge. I had wanted to drive you, but you went with Chet. When you called, you were all business. Normally you're very friendly. We talk. And then it's like a different Jessica shows up. You're all business, like you have tunnel vision. You look different. Your brow goes down a little bit."

He's smiling. A semi-smile. He's embarrassed.

"You were in the zone that time with Stevie," he repeats. "You were a little impatient. You didn't even notice that I was there, and I was feeling you didn't notice Chet was there. So you were functional at one thing. Just one thing."

"Did I ever hurt your feelings when I was like that?" I ask.

"No," he says.

Of course I want to hear whatever it is he has to say. But I also want to float out of the room, away from the intimacy of this embarrassing moment. I touch my clothes, making sure they are straight, making sure they are still there.

"No," he says again.

"Why not?" I ask, bluntly as usual.

"Because my dad was like that. I knew you were in the zone. I saw my dad in the zone so many times. You never knew who would come home from work. It could be the tough marine. Or was it the sweet guy who would read Robert Frost poems to you. When you have a dad like that, you learn how to read people's moods really quickly. You read their body language."

Again, I straighten my clothes, hoping he won't notice.

He continues. "I understand what you're doing when you interview terrorists, when you interviewed the people in this book," he continues. "You were reading these people. You are good at reading people. You know how to sense danger quickly."

How does he know this?

"I wasn't offended when I saw you in the zone. I found it endearing," he concludes. That word.

Jack has told me a number of times that he finds my various incapacities—my inability to maintain control of documents connected with researching this book, my tendency to get lost when I'm agitated—"endearing."

"I think people perceive you as tough. You've had this career, working for the National Security Council, working on terrorism. The things you've done seem very serious. People probably think you're an all-business type. I think the way you seem so tough has worked to your advantage. When you're in the zone, people are impressed by you, but they're also afraid of you. You don't seem approachable."

He sees that I am puzzled.

"Remember that guy in New York?" he says, referring to a decorated military officer who had wanted to talk to me after a speech I gave, a speech that had made me nervous, and presumably put me in the zone.

"He was a captain in the army!" Jack reminds me. "But I think he may have been afraid of you. He had been stationed in Afghanistan for a year and a half. But he was nervous about talking to you! He came up to me because he wanted to talk to you. He knew I had accompanied you. You didn't seem approachable."

He tells me that if I want to see how I come across when I'm in the zone, I should look at two different talks I gave, which were both filmed. "In the first one, you're laughing and joking with the audience. You look relaxed. In the second one, you look stern and very serious," he says. "That's what you look like when you're in the zone."

I have no desire to see myself in the zone. I don't know why. I don't look. Even after he reminds me to, a few weeks later.

"I had been working on this project for a year and a half, but I hadn't really thought about the connection to my own experience. But then I talked to Jen again. She was sitting at the bar. I told her, it's strange I'm doing this job for Jessica and I had this experience myself. And she was like, you have to tell Jessica this. I told Jen, My job is strictly to get the information that Jessica needs. She told me, If you don't tell her, I'm going to."

"Why was she so insistent that you tell me?" I ask.

"She thought it was our fate to work together. She thinks that way."

We are both slightly embarrassed by this sort of talk, the idea that fate could play a role in our work. But Jen is an artist, so it's okay.

"I have a confession," he says, smiling shyly.

How can he feel shy with me after all we've been through together? He has seen a part of me that I would have taken great pains to hide, had I known what I was revealing.

"I listened to the tape you gave me, the tape of you and your sister talking to the police after the rape. I couldn't stop. You sound young and you sound so bold and brave. You're laughing. You're not cowering in the corner. It's heartbreaking. It was traumatizing for me to listen to it. Because you're a kid. It's no longer easy to see you as a statistic when listening to that tape. It's over an hour long. No one was with you. Your parents weren't with you. What were you doing there alone, talking to the police without your parents? Describing the gun. Why was no one there with you? It felt voyeuristic, listening to the tape, but I had to listen. I felt dirty. Disgusted."

I don't mind that he listened to the tape. I'm relieved. What I do mind is this zone thing. No one has ever seen that before. At least, they've never told me that they did.

A few days later I, too, try to listen to that tape. I am in the car. At first all I hear is the sound of static. Jack had the tape remastered to enhance the sound, but a loud background buzz remains. Then I hear the sound of papers rustling, and the voices of two police officers. I recognize their accent and intonation—a style of speech that my sister and I used to call a Concord accent, which sounded so strange to us when we first moved away from New York. Then I hear a girl's voice. Is that my voice? Maybe it's my sister's. I feel uncertain. I stop the tape.

I need to pay attention to the road. I can't make out that girl's words in any case. A staticky buzz, the sound of artillery, supersedes her voice. No point listening. Plus, there is a thought, or maybe a sensation, pressing in against my head. A kind of aura. It's not real, of course. But it feels as real as the kaleidoscope pattern that blurs your mind when a migraine aura strikes. The feeling that my head might crack open like the shell of an egg, and I would fall in. An empty white eggshell. I can almost hear the cracking sound. I do not want to hear any sounds. I drive to my destination in silence and forget about the stupid tape. Until now.

"For me," Jack says, and pauses.

"You don't need to know this." He pauses again.

I hear a truck's brakes screech on Charles Street, a block away from my kitchen, where Jack and I are sitting, talking about him and about this work for the first time. I wait for him to tell me more about how hard this project has been for him. I feel a wave of heat rising up to my cheeks.

But actually he wants to tell me something else, something unrelated to the emotional difficulties of investigating a child rapist.

"Here I am, this night-school student in a prestigious university. I'm hired to work on terrorism," he begins.

My pulse returns to normal.

"But then a few months later you ask me to help you with this really personal thing. It's not mundane," he says, somewhat breathlessly, like the student that he is.

Well, that is true, I suppose. Not mundane.

"A lot of people live pretty mundane lives," he says. "Most grad students are intimidated by their professors. Most of them hate their jobs. They're afraid of their professors! But the protocol between us isn't really well established. I don't know how I'm supposed to relate to you because I know it's so difficult for you. There is professional distance, but there isn't. Maybe you maintain the distance more preciously because of trauma."

What distance? I wonder. I am not aware of deliberately maintaining distance, preciously or otherwise. He has seen a side of me that I would much prefer to keep hidden. But I can't, it seems. At least not from him.

But then I realize that the cold persona I slip into, the one he says might help me in my career, might actually frighten Jack a bit, even though he knows all about what he calls the zone, even though he knows me so well.

"There is a kind of embedded anxiety in this job," he continues. "I would sometimes find information that was difficult for you to hear. I was really conscious of how I put things. A lot of these characters would say sexist humiliating things. Stevie, for example. He called Abby a pig. She was a pig that Brian Beat liked to fuck, he said. I think I left that out when I told you about Abby. I didn't know what to do—I didn't want to be patronizing. I didn't want to be paternalistic. But I wanted to protect you from hearing that.

"I could tell you didn't like Stevie. I knew that you didn't like him. I didn't want you to prevent me from going there. He gave us a lot of leads. We found out about Abby and Mary and all this stuff about him living on the Cape. I was most worried about the possibility that you might not want me to keep doing the work. I didn't want you to tell me not to go."

"You're a born detective," I tell him. Just like me, I think to myself. I recently recommended Jack for a part-time job at the FBI, and the special agent in charge is pleased with his work.

"Was this healing for you in any way?" I ask.

"This conversation is healing. I am glad we're talking. I was confused. Sometimes you're really sweet to me . . . and then sometimes the other Jessica is there. I have experienced that before. It's really personal work," he concludes.

Does he mean that the subject matter is personal? Or does he mean this sharing of what normally goes unsaid? I am not sure that Jack understands what he is telling me about himself and the reason he understands this zone thing.

"Did it feel like you're contributing to an understanding of PTSD?" I ask.

"It's relevant to my past," he says, not really answering my question. "But until Jen pointed this out to me, I didn't realize it. Until then it was just work."

Now I want to know why Paul Macone helped me. I want to believe that he helped me because he thought it was the right thing to do. But how can I be sure of anything these days? What if he had other motives?

I decide the best way to find out is to ask him.

I drive out to the station. At last, I am able to drive myself. But still, I feel that ghost of shame.

"I had to reopen the case because a guy like that could still be out there," he says. I know this. He has told me this before. But it's not really what I want to know. I want to know why he continued to help me, even after we had identified the rapist and discovered he was dead.

One thing I fear is that he helped me because he knew Chet. Chet became a state representative at age twenty-one. He ran for office as a way to fulfill a college requirement. Then, astonishingly, he won, even though a Democrat had not won an election in our town since the Civil War. He was the youngest state rep ever elected. And then he became the head of the Democratic Party in the state. And then he became a state senator. And then he headed up Ways and Means. And then he became a U.S. congressman. He lost his last election in 1992, in part because of his support of the Cambodian refugees who had moved into a town in his district, the same town where my father now teaches. Now he is in the private sector. But still. Maybe Paul was influenced by the fact that he knew Chet had political connections. Or maybe just because he knew Chet's family.

"No!" he insists. "I didn't know that Chet would be coming with you when I asked you to come in. I didn't know you two were involved when I decided to reopen the case."

I want to believe this.

"How did you meet Chet?" I ask. He doesn't seem to understand that I'm trying to smoke him out.

"He was a little older than I. Somebody that everybody looked up to. He was friends with my brother," Paul explains, smiling as he recalls. "The Atkinses, they were a well-thought-of family. You mention Chet Atkins. He's a somebody."

"But Chet was . . ." He pauses, searching for the right word, still smiling. "He was a character. He once brought a live pig to school. Another time he brought a cow. He and my brother also managed to haul a car up to the roof of the high school. He was a well-known prankster. A character," he repeats. Another time, Chet put bullfrogs in the girls' toilets.

I am mostly satisfied, having watched Paul's face during this recitation. He is not pretending that he didn't know that Chet was a "somebody," which somehow makes his claim that he wasn't influenced by Chet's involvement more credible.

But I can't quite believe him yet.

"Why did you become a police officer?" I ask, trying to put the story into context.

"Lenny Weatherbee and I were old friends. [Lenny is now the chief of police.] We went to high school together. I was teaching shop at a special-needs school at the time. In Natick. I was teaching industrial arts," he says. "Auto mechanics, photography, and machine shop."

He sees that these words don't resonate with me. "All the things they taught us in the corner of the high school that you probably didn't go into.

"Lenny had just come onto the force. I used to see him here in Concord. One day, I remember we were at my girlfriend's house at the time. She's my wife now. He said to me, 'Why don't you come on as a Special. It will be fun. We can ride around and be cops.'"

"What does it mean to be a Special?" I ask.

"It's a part-time job. They give you a gun."

"Did you know how to shoot it?" I ask, astonished that a person trained to teach at a special-needs school would be allowed to work as a part-time cop. I have shifted into my interviewer mode, the shame shimmering, almost invisibly, in the distance now.

"Back then we had shooting in gym! Twenty-two-caliber rifles. Don't you remember?" he asks.

I don't. I do remember that they would let us out of gym if we were studying ballet, an enormous relief to me. Learning to shoot might have been more useful to me than ballet, as it turned out.

"You had to get a license to carry a weapon. I bought my own gun, a Smith and Wesson .38 caliber. You had to take a test in basic criminal law. We don't have Specials anymore," he says. "Accreditation reasons. Liability reasons. This was 1978.

"So my friend was on the force. I thought, Why not? I wasn't married yet. I was still teaching, but you could do it at night. Lenny was a patrol officer. The lowest rank. We went out together in the cruiser."

"Did you hang out at Brigham's?" I ask. In our small town, it always seemed that the cops spent most of their time drinking coffee. There was so little crime.

"We went to Friendly's," he says. My heart sinks. There were only two places you could get coffee back then. Brigham's had better ice cream. Friendly's, I knew, had a pay phone that worked.

I've been in this station before, I am certain of this now. It was long ago. Paul doesn't seem to sense the aura of shame around me, or maybe he's just too polite to say. I've hidden this shame for so long, with security clearances and certificates and degrees.

"What sort of crimes did you have to deal with?" I ask, wondering how a teacher would know how to respond to a crime. (After all, I'm a teacher now, too).

Most of what you do in a small town like Concord is enforcement of motor vehicle laws, Paul tells me. There was the rare case of breaking and entering, he says. "Back then, on the weekends, kids were always having parties, and we'd have to tell them to be quiet. Sometimes they had parties in the woods. I remember one party at Ruggiero's Piggery. A lot of kids smoking pot and drinking."

"But you weren't much older than the kids!" I say. He was very straight back then, he tells me.

"Were you excited to be carrying a gun?" I ask, trying to place him in a context I might understand. I'm accustomed to talking to men who like guns, accustomed to pretending it doesn't frighten me. I still cannot bear the thought that Paul might have been acting out of goodness, out of kindness to me.

"The first time you wear a gun, you're bad."

I'm not entirely sure what he means by this expression, but I'm relieved that he told me the truth. And relieved that this "badness" was in the past.

He must have noticed my confusion. "You feel pretty important," he explains. "It lasts about a day."

"Did you ever have to use a gun?" I ask.

"To this day I've never shot at someone. . . . I've been a cop for thirty years."

"Did you ever have to threaten anyone?" I ask.

"Never had to threaten. Very, very seldom in Concord does anyone on the force have to use a gun."

"Then what?" I ask.

"A job came up. I became a patrol officer. Bottom of the barrel. I used to see the detectives coming in every morning, wearing sport coats, carrying their cups of coffee. You come in grimy. You've just finished riding around for eight and a half hours, all night long. From the very beginning it seemed like being a detective would be more fun. More challenging intellectually. I come from a mechanically oriented family. When you're a mechanic, you're curious about how things work. You know the engine runs well. Why? You take it apart. Same thing when it breaks down. You take it apart. You're curious."

It turned out that Paul Macone was a talented detective, and he rose to be second in command of the force.

Now I want to test out another troubling theory. I'm still not ready to believe Paul.

"Was it because you know I had worked in government?" I ask.

"The first time you came out, I Googled you," he admits. "I was a senior in high school the year you were raped. I was always in the machine shop, that's why we didn't know each other. Whenever someone asks for a file, I try to figure out who they are, why they're asking. You have to be very careful with a sexual assault case. You don't want to retraumatize a victim. I saw you worked on terrorism, that you had worked in the Clinton administration. I could see you were someone I didn't need to mince words with. Terrorism sometimes equals death. Terrorists do terrible, awful things. I thought to myself, This is not an extremely fragile person. Knowing the kind of work you had done did make me feel more comfortable sharing information with you," he admits. "And you seemed sincere. You seemed like you wanted to understand who your rapist was for the right reasons."

What would the wong reasons be? To kill him?

"And then it turned out you were able to help me." He smiles. "If I noticed something, you'd bring up a counterpoint. It was helpful. You were looking at the case with me, acting like a colleague even."

And now, finally, I remember. I remember the source of this shame I feel when I am in the police station. I'd been summoned into that station before, several times. My sister and I had to look at photographs of suspects in our case. There were none we could recognize. I must have sensed that the police thought that I was lying, that the rapist was, as they told my father they suspected, someone that I knew.

Victims, I somehow "knew" this to be fact, are ineffectual, weak, and dishonest persons who drag society down. I understood that terror and despair were contagious emotions, and that to indulge oneself in the feeling of terror was antisocial and possibly even immoral.

I must have wondered if the police were right, if the entire story was a figment of my imagination. This is the worst impact of severe trauma: the victim loses faith in the evidence of her own senses. And this is the great gift Paul Macone gave me. He believed what I told the police back then. He believed me enough to try to solve the case,, and he did.

Perhaps because I've sought out evil in this world, attempting to understand and tame it, I am particularly moved by goodness. There is a light that animates an act of generosity, when a person is kind—not to call attention to his own goodness, or to make a pact with God, but just because he feels it's right. I see this light in Paul Macone. Still, his kindness is almost too much to bear. I feel shy around him, despite this conversation. I even feel shy writing this down.

chapter ten

Collateral Damage

The spy in me cannot stop pondering the meaning of the stones that Detective George Remas found in Brian Beat's front pocket, as well as the stones that Beat placed nearby some of the girls when he raped them. I look into Detective Remas's claims that there had been a series of pedophile priests at the church that the Beat family attended and discover that they are true. But those priests arrived after Brian Beat grew up. Could pedophile priests have been placed there routinely?

I hear rumors about sexual abuse at Beat's elementary school, during the period he was there. Chet drives me to visit the school and we speak to the priest who runs the school now. There may have been abuse back when Beat was a student, but there is nothing in the school's files that would indicate there was, the priest tells us. Then I learn that the region where Beat grew up was a dumping ground for pedophiles. I decide to contact one of the victims of clergy sexual abuse in the area who has become an activist. His name is Skip Shea. The priests who abused him were from a town located next to Milbridge, Massachusetts, where Beat grew up.

It ought to be a lot easier to talk with victims than with killers or rapists. They're in pain, they're afraid, they've been wronged. I've studied terrorism for more than twenty years, and I've interviewed hundreds of terrorists, but I've never interviewed a victim. Before commencing this book, I'd never talked to a victim of rape. My answer to the obvious question, "Why not?" is that I already know what it feels like to have a gun trained on you, to fear that someone is about to kill you. But actually, I'm not sure this statement is true. I had these experiences, but I'm not sure that I felt them. Perhaps I should just admit it: I'm afraid to empathize with a victim. I am afraid.

There must be a frequency that victims acquire. I am afraid that the experience of being in a room with Skip will shake off a facade that protects me and holds me together—to reveal not only my fear but also my rage and shame.

I'm trying to get this right, to explain my embarrassing position. Let me try this once again. I cannot bear to be around victims who see themselves as victims. I'm more comfortable talking to victims who are numb, or who have learned how to harness their unfelt rage and fear to do productive work. I'm most at ease with the sort of victim who ends up doing work that involves exposing himself to risk or violence—soldiers or human rights workers who work in danger zones, whose love for humanity is expressed without a display of feeling. But Skip is not this sort of survivor. His life's work is to help other victims of clergy sexual abuse, which requires him to feel his own abuse, again and again, on a daily basis. This is what is hard for me to witness.

• • •

I am relieved when Skip tells me that he will come into Cambridge to see me; I don't like driving out to the area where Brian Beat grew up. For me, Milbridge smells of evil. For me, terror smells of cheap cologne and air purifiers, like the air purifier in Mary's bathroom. And it's only now, having just written that sentence, that I recall my observation to the police that my rapist wore cologne. For Skip, evil smells of cigarettes and Southern Comfort—the scents of the first priest who abused him.

"When were you first abused by a priest?" I ask.

"When I was eleven. Father Billing brought me into the basement of Saint Mary's Church, near the nurse's office. We were goofing around. He reached for my pants. I was terrified. He talked about love. I was an altar boy. He was an authority figure. I did as I was told. It continued, for months."

"Did it happen every day?" I ask.

"Often there would be a couple of weeks between incidents. I was an altar boy, I attended his masses, so I saw him often. It escalated. He began taking me out of my CCD [CCD stands for Confraternity of Christian Doctrine, the religious education program of the Catholic church] classes—he would take me behind the rectory. He told me, 'This is love, this is all about love.' I cried. He would say to me, 'God doesn't want to see you cry.' So I learned not to cry. Since then, I've had a hard time crying."

I know all about this—the difficulty that abusers of children have dealing with the impact of their abuse, how they try to brainwash their victims into thinking that the abuse doesn't hurt, that it is actually good for them. If the abuse is sustained over time, the victim learns not to feel. But I don't tell Skip this. I listen.

"Did you tell anyone what was going on at the time?" I ask, gently, I hope.

"No. I was afraid to. I grew up in a very religious home. In a very Catholic town. There were pictures of priests hanging on the walls of our kitchen. In our house, priests were next to God. I assumed that no one would believe me."

"How did your family react when you finally told them? Did they believe you?"

"Yes. But I didn't tell them until Billing had already been charged. He had been moved out of our church some time before that. I was in a suicide shelter.

"But he was not excommunicated," Skip adds.

We are sitting in my living room, directly across from each other. I am sitting cross-legged, my computer in my lap, slumping under the weight of the shame that hangs in the air between us, the shame we both feel. Is shame contagious? I believe it is. But now there is a new energy in the room. Rage. This part of Skip's story is still undigested, a bitter portion. The state has taken Billing off the street. The church has forbidden him from preaching. But the church apparently does not view the sexual abuse of children as a sin sufficiently serious to warrant excommunication. These are the crimes that the church considers to be the most serious sins: Attempting to absolve a person who has committed adultery. Acquiring an abortion. Violating the confidentiality of confession. Physically harming the pope. But not repeatedly persuading a child that allowing himself to be sodomized by a priest is an act of love.

"There was a bishop in Texas who wanted Father Billing to move out there," Skip continues. "He knew what Billing had been accused of, but he still wrote a letter asking for Billing to be sent to his community. You can read that letter yourself," he says, telling me where I can find it.

"Father Owen replaced Father Billing when Billing was removed from our church. Billing owned a house down on the Cape, and he and Father Owen used to go down there. Later, I realized, they must have talked about me. Owen replaced Billing in every possible way, including abusing me."

Why does everyone in this story seem to be spending so much time on the Cape?

"Father Owen took over for Father Billing when I was thirteen or fourteen. He was more appealing than Billing. He had a really cool stereo," he adds.

"Where did he try to have sex with you?" I ask.

"In his room, in the rectory. He was a teacher. He regularly brought students into his rooms, his office was there."

"How long did Owen abuse you?"

"Until I was sixteen or seventeen."

I am drawn by cruel thoughts. Not when I was with Skip, but now, reading over these notes. Unseemly, embarrassing thoughts that I would normally shoo away, and would normally not want to admit, even to myself. But now I wonder whether others have had similar thoughts about me. Perhaps it is only victims who hate other victims. Perhaps victims who still deny their terror.

Skip was immobilized by terror. How unbecoming to feel afraid. I want to condemn him, first for feeling afraid, and second for not running away. I want to condemn him, in particular, for allowing his victimization to become such a central part of who he is. He calls himself a survivor, not a victim. But "survivor," it seems to this cruel, censorious side of me, is just a faddish, politically correct way of saying "victim." Is this why my father refused to accept reparations from Germany, because it would infuse "survivor" or "victim" onto his identity? Is that why my father reassured the police, several months after my sister and I were raped, that we had gotten over it—that he didn't want his children to take on the mantle of victim?

What is the difference between a victim and a survivor? Survivors, we are told, emphasize their own agency, and are thus different from victims, who cringe under fate's blows, passively accepting fate's amoral and senseless punishments. But what agency did I exercise in my own survival? I stood still, entranced by my rapist's gun. I made no quick movements. I did not grab the gun from the pillow and aim it at my rapist. Why? Because I was immobilized by terror. And I wonder, too: What agency did my relatives exercise in their own survival—not the ones who fled, but the ones who remained behind in Germany, ultimately to perish? Were they, too, frozen by fear?

Freeze, fight, flight. Freezing is the first reaction. But a person can get stuck, frozen forever. Flight is not an option for a slow-moving animal facing a gun, slow-moving animals such as girls. And flight is not really an option for boys who believe themselves to be serving God by servicing sick priests. Victims, unless they are trained, do not get to choose the way they will react when they are "scared to death," the phrase we use to describe the altered state that is evoked when someone or something threatens, credibly, to annihilate us, body or soul. Body or soul or both. Skip and I have this in common: we both froze, and we are both still immobilized, at least some of the time, by shame.

"Wouldn't you have been able to overpower him by that time?" I ask.

Did I press him too much? But there is something shocking about this—a teenage boy unable to fight off a priest.

Skip has big biceps, I see. I imagine his thick arms snapping the spindly legs off his predator priest. In my fantasy, I join him. I have strong arms, too. To join him in fighting the supernatural power that was attached to a sickly specimen of a human being who held himself above others by virtue of his supposed connection with the divine.

What is a victim? If I keep thinking about victims, embarrassing thoughts rise to consciousness. Embarrassingly cruel thoughts that are not politically correct. One feels sorry for victims, but one also feels lucky not to be one—even a bit superior, a not entirely unpleasant feeling.

Victims are weak. They must be. Why else would they be victimized? Especially rape victims. Morally and physically weak. And then there is the comforting thought that victims lie. If the victim is exaggerating or has made his story up out of whole cloth, I don't have to confront my own unattractive desire to punish him. For the first time in my life, at age fifty, I realize that I might feel defiled by what the Nazis did to my relatives, the way my father and his brothers were forced to flee in terror, leaving their home behind; the way my father's cousins were murdered.

I hate victims because I was raised, from early childhood, on this prejudice. I know it's wrong to hate victims, it's ugly, I'm ashamed of it. Yes—I blame victims. Yes, I blame myself—both for having been victimized, and for hating victims. Yes, I know, in the sane part of myself, that it is wrong—both morally and logically—to blame victims.

And all this leads me to wonder about you—my reader. Are you feeling sorry for me, but also a bit superior? Are you toying with the idea of imagining that I exaggerate, that I made any part of this story up? I will reveal the secret perpetrator within me—the secret perpetrator hidden in every victim of violent crime or sexual abuse: I'd like to see you try getting through what I got through. I'd like to see how long you could stand, how long you could sustain both sanity *and* humanity.

"I was small, and he was huge," Skip explains, responding to my question about why he didn't attack his perpetrator. This perpetrator, in my mind's eye, is a broken man with skinny, easily broken limbs. I have made the mistake of imagining that the strength of the priest's body matched the weakness of his spirit.

"But I finally had the courage to demand ending it because I became very interested in a girl," he says, surprisingly.

"How did Owen react when you told him this?"

"He said, That's okay, because I was going to end it anyway. I was going to end it, he said—and this is the worst—because you're inadequate in technique and in size.

"You're inadequate in technique and in size," he repeats. "Those words have haunted me ever since. After that I couldn't have a relationship with anyone that wasn't initiated by them. So much shame. Ashamed of what I did, and ashamed of my body."

This is what occurs to me now. I would like to kill this priest. At the moment, the thought of violence against pedophiles is extremely appealing.

There is something attractive about the idea of becoming a terrorist in response to being terrorized. One would like to respond to terror in kind—to maul or perhaps kill one's assailant in self-defense. I could smash this pedophile. I could explain to him, "You're inadequate in body and soul. You're not a real man."

I'd like to have murdered my own rapist, too. But physically defending my sister or myself is not all that I would like to have done. I would like to have terrorized my perpetrator, to have returned terror for terror. A court of law might well exonerate me for violent actions taken to defend my sister or myself from harm. But would a court of ethics forgive the hatred I feel, the desire not just to defend my sister and myself from harm but also to terrorize and to hurt our rapist? And what if I, unable to terrorize my own perpetrator, turned my rage against others? What if I became a professional terrorist? Would my rape, or the assaults my family sustained, be sufficient to excuse, morally if not legally, my own violence against others? The answer, I believe, is absolutely not.

"First Father Billing, then Father Owen, then a third priest facilitated by Father Curran. I don't know the name of the third one. Again and again. I don't think my story is that unusual in the church. It was a common thing to pass kids along from priest to priest. Pedophile priests formed rings. Billing was also involved in a ring in Texas after he was pushed out of the ring around the House of Affirmation up here."

"Do you know how they selected kids?"

He isn't sure, he says, but then offers an idea. "Pedophiles go after victims who have just suffered a major trauma. The priest comes in to help the family, and then abuses a child. They also go after families with a lot of faith—faith itself can be a kind of weakness in their eyes. They exploit individuals who have suffered severe trauma or who have strong faith," he summarizes.

"My drinking started with Father Billing," he says, turning to a new topic. "Alcohol was a big part of this. The first time he gave me a drink in his car. It was sweet, Kahlúa . . . which helped immensely. It made the moment tolerable.

"These are holy men. I started to think it was my job to help these priests, who had God's job to do. I can't even say that I was surprised when the second priest seemed to pick up where Father Billing left off. In the Catholic world, there is a lot of mystery, a lot of mysticism. I thought it was just my role to help them, to serve them.

"There was terrible collateral damage from all this," Skip adds. "The main collateral damage is that I drank, and I hurt my children deeply. But I also started seeing shadows of people who weren't there. And smelling things."

What kind of shadows?

"I don't know," he says. "A hooded figure."

A flashback, I think to myself.

"Is it a shadow of a priest?"

"It was always dark. . . . I think it was a priest. There was a hood or a hat. . . . It was like a poster for *The Exorcist*—a priest walking through the fog."

"Did the priest ever talk to you?"

"No."

"What did you smell?"

"Sometimes I would smell cigarettes."

"Did Father Billing smoke?"

"Yes. Sometimes I would smell liquor—Southern Comfort. At first I thought I had drunk Southern Comfort at my home, that I was remembering something that happened at my parents' house. But my parents didn't drink Southern Comfort; Father Billing did, and I would smell it on his breath. He also got me to drink it, to make me more relaxed during the sex act.

"I know I'm under stress when I see the shadows. These shadows come to me when I feel unbearable anxiety."

"What are you anxious about?" I ask.

"I am most anxious when there are stories in the paper, and people refuting the story of sexual abuse in the church. I am anxious about people knowing what happened to me. People thinking I was gay. Sometimes I don't know—would it be a bad thing to be gay? I feel terrible shame.

"So my children had a drunk for a father. The shame is huge."

I take this in, what it must feel like to recognize the harm we do to our children as a result of all the ways we strive to avoid feeling our own pain. I am distracted by a recurrent thought. I worry that if I don't feel the pain of my mother's death and the terror of rape, I will harm my son, Evan, in some way, perhaps by overprotecting him. But I need to stay in the room. I will feel about this later.

"The collateral damage, it goes on and on. . . . I taught my twin girls pro wrestling techniques. I wanted them to be able to protect themselves. I thought it was a good thing to teach them this. I realized, much later, what I was doing—teaching them to defend themselves from potential pedophiles and rapists." Just as my father, I believe, was training us to survive a war.

"When does it end—this collateral damage? It gets passed on and on, from one abused child to the next," he says.

• • •

This is my hypothesis. Terrorizing others—including by raping them—is a way to reassert one's manhood in the face of extreme humiliation. Feeling terrorized is humiliating. Having been raped is humiliating. To be treated "like a woman" is humiliating. Thus, the lament of one of the victims of sexual torture at Abu Ghraib, "They were treating us like women."[1] Rape is a perfect way to discharge one's shame. But like fear, shame is contagious. The shame and fear of the rapist now infect the victim, who, depending on his psychological and moral resilience, may discharge his fear and shame into a new victim, not necessarily through rape. I do not mean to assert that all terrorizers have been humiliated, or that all people who are severely shamed will ultimately terrorize others. My hypothesis is that shame is an important risk factor for savagery.

"I often feel like nobody," Skip says. "I ask myself: Why would you want to talk to me? Why would anyone want to talk to me? It comes on me suddenly, this feeling that I'm not anything . . . a person who has spent a lot of time in bed, who doesn't want to be anything."

I know what he is talking about, and this time, I tell him that. For years, I could not understand why anyone took me seriously. I could not understand how I managed to get into MIT or Harvard, why anyone would offer me a postdoctoral fellowship or a job. I could not understand why people kept turning to me after September 11. I didn't see myself as a person who couldn't get out of bed, but as a salesgirl in a coffee shop—the job I had as a teenager who was afraid to apply to college. My identity was stuck there for years.

"Inside me, there is the person who wants to be dead," he

says. "I can't advocate for myself. I can advocate very strongly for others, but not for myself. When I realized that my work might prevent additional suicides, I began advocating strongly for victims of clergy abuse, but it's much harder for me to do this for myself. Sometimes I'm not even sure that I exist. Is this really me—this person whom people want to consult about clergy sexual abuse? Or am I really the person who can't get out of bed? I've gotten better—I spend more of my time living in the present. But it takes a lot of effort to stay in the present—a lot of yoga and meditation."

I ask him whether he goes into altered states. I ask him for myself. I have read that other people experience the kind of altered states that I do. But it's another matter to hear about it from a live person. I want to hear what it *feels* like.

"The most annoying dissociative state that I experience is when I feel very sleepy. It is not really safe for me to drive. Have you ever felt that one?" I ask him.

"The sleepy one. Yes. It feels like eyelids half closed, feet in mud, slow motion. My brain is in a total fog. In this state I feel completely unsafe.

"Sometimes," he adds, "I become a prisoner of detail. I see too much, hear too much, it is painful. Hypervigilance. Like a cocaine high. The mind is hyperfocused, heightened awareness. I have an ability to scan quickly. . . . I feel super-smart. I seem to be able to problem-solve a lot quicker. If someone is looking for something, I can find things. . . . I could do a jigsaw puzzle ten times faster in that state. It is almost as though I have hypervision. It is not a state I mind being in. I can sense the feeling in a room. I can sense the mood of a crowd. If there is aggression in a crowd, I can tell just by scanning the place and stepping inside. I have a feeling. It's as if a veil were lifted—you can feel people's feelings. I can sense when there is sexuality in the air. I can be incredibly attracted to it, or I want to run. . . .

"I can use these states to my benefit," he concludes.

I know what he is talking about. There is a state where I feel smarter than normal. That often happens when I feel endangered, or I fear that someone I love is in danger. I agree that this state is not entirely unpleasant, and it is extremely useful when someone's life is in danger. It is also useful for highly stressful work. But the stress is addictive. I believe that this "high," as Skip puts it, is what makes some people stay in dangerous jobs.

When I am truly in danger, my reaction time is very fast. I notice everything around me, and think and act quickly. If I feel agitated but there is no physical danger, my mind will sometimes focus on unimportant details. And I, too, feel unsafe when I'm in the sleepy state that Skip describes.

I want to learn more about these states. Soldiers, I'm told, suffer from them as well.

I ask Skip what has helped him the most. "Meditation," he said. "Also yoga. Anything that keeps me in the present."

chapter eleven

Child Victims

I decide that I want to read what the other girls in the neighboring towns had said about their rapist, to see for myself how similar their stories were to mine.

The town counsel in Concord gave the police department permission to release redacted material from the files of rapes that involved a similar MO to mine. These rapes occurred between 1970 and 1973, in Concord and nearby towns.

I walk over to my trash can, the one with the elephants. I see a packet of files, held together with one of those jaunty-looking paper clips, yellow with red stripes.

I look inside one of the files. I see the girls' careful penmanship. They were writing for the police.

I realize, once again, that I can't read the files at home. I ask Chet to take the trash can and put it in the trunk of my car. I will take the trash can to another state and read the contents there. Chet often lifts heavy things for me when I'm traveling. The trash can is not heavy. But, anyway.

I wonder aloud, Would it make sense to transfer the files to a bag that is easier to transport? Perhaps this one here, a small black bag on which is written "TrueValue Hardware"?

"Let's leave them in the wastebasket," Chet responds. I do not ask him why. I think to myself, Perhaps he thinks the files are safer in the trash can. Perhaps he thinks that I am safer with the files in the trash.

I take the trash can from the trunk of the car into the cabin where I'm staying. I pull out a few pages. I begin to read. I see the words *bed* and *tall lean man with a gun pointed at me.*

I stand up. I am cold. Maybe I should make a fire. Once my kindling is crackling, I add a small log, bark side down. The bark ignites.

I sit down to read a page. I see the words *grayish black pistol,* and *he said don't scream,* and *do as you're told and you won't get hurt.*

I decide to make a pot of tea. I fill the pot and straighten up my desk. By the time the water boils, my fire has gone out. I insert fresh kindling. I push some logs around.

Once again I sit down to read the file, with a cup of tea warming my hand. I read a few more sentences.

I see the words *bend my knees wide open* and *unzippered.*

I get up. I am not upset. I just can't sit still.

Finally I decide to bring the trash can somewhere else. I take it to a nearby library, where there are people, and I have reason to hope I will be less likely to feel spooked.

I find that I must think carefully about where to put the trash can. I do not want it in my line of sight, nor do I want it behind my back. I find a spot just outside my line of vision, on a forward diagonal to my left, on a black leather chair. I am not a superstitious person. I feel embarrassed, even in front of myself, that all this seems to be necessary.

I have read about triggers—the sounds or smells or sights that can suddenly catapult us back to the moment when we faced a violent death; the way the sound of a backfiring car can bring a grown man to the ground. But it is hard to understand how this works. The soldier knows that cars will backfire. Can't he learn to ignore the sound? Can't he use his rational mind to overcome his response to that sound? And can't I, similarly, knowing that these files pertain to a dead man, treat them like ordinary paper?

Finally, sitting in the library, I am able to read through the files. I am not allowed to quote from the redacted reports, but here is a summary.

The rape at Concord Academy occurred at 11:30 PM on April 29, 1971. A tall, lean man walked into a student's room, wielding a gun. She was sitting on her bed, doing her homework. He pointed a gun at her. She observed that his eyes were a clear whitish blue. She described the gun as a gray-black pistol with a white handle. He told her to do what she was told and she wouldn't get hurt. He asked her how old she was, and whether the girl in the next room was pretty. He told her to put on a skirt and to dance for him. After that he directed her to the bed, where he made her bend her knees wide open and began touching her "lower parts" with his hands. He put down his gun on the condition that she wouldn't scream. He also put two rocks down by the gun. He pulled out a gray tube. He rubbed a salve on her "lower parts" and said it would make it slipperier. He tried to have intercourse with her but did not succeed. He asked her to go into the neighboring girl's room and wake her neighbor up, and to make sure the second girl wouldn't scream. The first girl reports that she opened the door and called softly to the girl in the neighboring room, telling her to be quiet, promising that the man with the gun would not hurt her. The second girl screamed in fright. The first girl reports that she muffled the mouth of the second girl to keep her quiet. The second girl calmed down. Then the man raped the second girl. The first girl reports that she heard the second girl's cry of pain. As he left, he told the two girls not to call the police. He also told them it was a cap gun.

I know these files cannot hurt me, but I feel them to be dangerous. As I read, I sense my body slowing down, as if I were dying. How could it be that this man managed to terrify a girl to the point that she was willing to "muffle" another girl's mouth? How could it be that the second girl calmed down, with the weight of the first girl's palm on her mouth? I want to scream.

I want to be precise here. I know I am not dying, or at least if I am dying, it is only in the sense that every second we live we are one second closer to death. But the weight of the words drags me down until I feel myself bent down like a little old lady in a fairy tale.

I get up, of course. I get up again and again; I cannot sit this through.

This is the part where my breath gets caught—the way the first girl talked the second into complying.

Why did she tell the other girl to be quiet?

Why did she promise the second girl that she would not be hurt?

Why, when the second girl didn't obey the command to be quiet, did the first girl muffle the second girl's mouth?

This is how the Nazis got the Jewish elders to arrange the orderly deportation from the ghettos to the death camps, how they got the Kapos to do their dirty work. Scared them half to death, but then let them live. In the moment a person is broken by terror, she is more easily seduced into "muffling" others. You will move up in the world if you only follow orders, if you sell out your sister, if you muzzle her.

Again and again, my mind gets snagged on "muffled." What does it mean? A muffled scream; a light snuffed out. To muffle means to suppress, repress; to smother, squash, or kill. This is what most rape victims do, they muffle themselves. I think again of the Muselmänner, the "walking corpses" in the concentration camps, who lost the will to live. They were not sent to the gas chambers, but they died, nonetheless, from terror, usually within days of arriving at the camps.[2]

Their light was muffled. You could see them, but they would not see you, Primo Levi tells us. They did not bother to avoid blows or to seek food, he says. The other prisoners could easily identify these condemned men as "Muselmänner," or "the drowned." The other prisoners steered clear of them, as if their suicidal despair were contagious.

This is what muffled means to me. Muffled is the way I let go, let myself down into my own nonbeing, when I thought the rapist was about to kill my sister and me; or perhaps my sister or me. I want to emit the terror that was muffled back then. I want to bellow like a creature facing slaughter, on behalf of my sister and me and on behalf of these two girls, the one attempting to stifle the other and the one struck dumb by the other girl's palm.

I consider walking out of my cabin and into the woods and screaming as loud as I can. This time I will not be muffled: I will lacerate the ear of God. I will wail wildly: How could you let this happen? I will scream so loud that the molecular structure of the air will change. You have not heard of this happening, you will say. But just wait and see. I will roar argon into chlorine, xenon into fluorine, all the noble gases into reactive ones. My lament will terrify even the stars.

I see, in my mind's eye, the tremor of the stars as my dirge reverberates into space. I will shout out the sound that has been building in all of Brian Beat's victims for decades.

But what if my scream isn't loud enough? What if the stars don't tremble? Even worse, what if they do?

I could go out in the woods and scream, but I'm too inhibited. It seems that I muffle myself. Or maybe I'm afraid to wake the dead. If I truly rage, if I truly grieve, Brian Beat would surely hear me, dead though he is. He might come back and get me. I am afraid that if I scream too loud, this time he will kill me.

Is this the reason I lost my capacity to feel afraid? I was too afraid, perhaps. Irrationally afraid. Once it starts, it might not stop.

I get up again and again. And yet I feel a kind of duty to read the raped girls' words, so I read the second girl's story, too.

She was asleep when the girl in the neighboring room came in, followed by a man with a gun. The first girl told her, she said, not to scream and to do everything the man said, and then promised her that she wouldn't get hurt. It was because she trusted her neighbor, she said, that she complied. Then the man with the gun commanded her to dance. He spread the salve on her. When he entered her, he admonished her to loosen up, she was too tense. He had clear blue eyes, she wrote, and was tall and thin.

There it is again; the first girl muffled the second one. Even so, the second girl trusted the first girl. Be quiet, the first girl said, so that we can both survive.

I've been quiet, too, for the same reason. I've been quiet for years.

The Muselmänner disintegrated into a state that "signaled the approach of definitive indifference," after which there was no turning back.[3] Physical death was imminent, but the Nazis did not kill these victims with gas. They killed them with terror.

But I'm not quiet anymore. No longer a Muselmann. What my

life tells me is that it is possible to survive, outwardly, while remaining blanked inside. To be a Muselmann only inside.

And now I wonder: Which girl suffered more—the girl who was penetrated, or the girl who wasn't? The girl whose cries were muffled, or the girl who has to live with the knowledge that she muffled her sister as a service to their torturer? The Jews who were killed, or the Jews who survived as a result of their work for the Nazis?

There are more files of similar rapes, the others at other private girls' schools in the Boston area. I read these files as quickly as I can. The rapist described in these cases is always slender and tall. Most of the victims described his eyes as "startling bright blue," "whitish blue," or "clear blue." He always asks his victims' age and where they are from. He always threatens to kill them. He always commands that they look away, but only dons a homemade mask during the act of rape. He does not seem to be deterred by the presence of potential witnesses: he often entered dorms or homes that were filled with people. There is almost always more than one victim, usually two; and in many cases, only one of the victims is penetrated. He seems to take his time. He likes to have an audience. He is particular about his victims' positions. In some cases, the victims notice a white handle or white grips on a gray-black gun. In other cases they describe the gun as grayish black. In almost every case, when the rapist is done, he admits that the "pistol" they observed was a cap gun. Some of the victims observed acne scars on his neck and face. He is usually described as having a "Boston accent," or "local accent."

His technique changes over time. In the beginning, he carried a tube of jelly, which he spread on the victim's vagina. He also carried stones in his pocket, and laid them out on the bed or nearby on the floor, as if reenacting some kind of ritual. Sometimes, early on, he wore a wig or gloves, but later he did not. He claimed to be from different places, often the state his victim was from, and he used different names.

It is a bleak and cold January day. My son is with his father, and Chet is away. I have been writing all day. My neighbor, Pebble Gifford, has invited me to her house for dinner, where there will be other neighbors. I have written too long into the afternoon, trying to take advantage of the silence in the apartment. It is hard to transition from the dead Brian Beat to the living world, but I take a walk and a bath and I find a way back.

It is a convivial Cambridge party. Most of the attendees are a generation older than I. They are sophisticated—well traveled and well read. Many of them have known each other for decades. They have worked on failed campaigns together. They've seen one another through triumphs and then a slowing down. It is a group to which I don't belong.

But I feel welcome. I also feel safe.

Two of us are outsiders in this group—Joe Finder, a successful writer of thrillers, and myself. At one point, our hostess asks the two of us to tell the group what we are presently writing.

I have told very few people the subject of this work. Not that it's a secret—it just feels odd to talk about it. I propose to Joe, the famous writer, that he go first.

But he, the famous writer, wants to give the floor to me.

It's not that I'm afraid. Of course not. What is the point of writing a book if I can't reveal its subject, even to a friendly audience?

I tell myself I have no colleagues here, that I have nothing to lose, that it's good practice to talk about the subject of this work.

"I am writing about my own rapist," I blurt out, perhaps a bit too quickly, perhaps a bit too quietly.

I sense confusion. There is a silence, but also a rustling, as people shift their bodies. They are uncomfortable with what they couldn't quite hear. Maybe they are trying to get a grip on the meaning of the words they heard.

One person asks, "What did you say?"

I start again. I tell the group that the rape occurred more than thirty years earlier, that the police reopened the case, that they believe they solved the crime. I tell them about the forty-four known victims, all between the ages of nine and nineteen years old. I tell them that twenty of the victims had been living in the eight-block area centered on the Radcliffe dorms.

I tell myself, I've done it. And even survived the telling.

One of the women, a widow, has been notably quiet until now. Vera. I know who she is. We all do. She is a longtime political activist who works on behalf of children. Her husband had been a famous professor of history, known throughout the world.

Now she speaks.

"One of the victims was my stepdaughter," she says, flatly.

I look at Vera. I observe her precisely cut gray hair, thick gray brows over deep brown eyes. I take note of her woolen suit, a sensible gray-green. Eileen Fisher, I think to myself. She could almost be in one of those ads featuring middle-aged women who look like they've lived enviably complicated lives, but are now nicely pulled together, thanks to their clothing designer. She is aging well, I observe.

She is thought to be kind. She is thought to be reliable. I know this. But she is approaching eighty. In her era, the world was smaller. In her world, rapes were rare; and rapes at gunpoint, rarer still. In her world, you could assume that there could not possibly be more than one rapist active in town in any given year. Therefore, it stood to reason. But I couldn't believe this. I live in a more dangerous world, one where rapists are more common, or so I imagine.

"I will contact my stepdaughter," she says, "and ask if she'd like to talk to you."

"Thank you," I say. I would like to talk to her stepdaughter. I've never talked to another victim of rape. But I'm sure, of course, that we were raped by different men.

Lucy, the stepdaughter, and I agree to meet. She will bring her sister as support, and I will bring Chet. I suggest Rialto, the best restaurant in Cambridge. I order the most expensive bottle of wine I've ever bought, in the spirit of the *Titanic*.

Lucy is almost exactly my age. Lucy's parents had divorced a few years before she was raped, just as mine were. Her mother died, just like mine; only Lucy was in her twenties when her mother passed away. Like her father, she is an academic. I know, before agreeing to meet her, that she is a successful academic. I hate to admit it, but I would not want to meet with her if she had been undone by this crime, if she identified herself as a Victim. She is a professor of criminology at the University of Vermont. She was raped at age thirteen, in her mother's home. The house was filled with women, her three sisters and a number of their friends.

She tells me that in the last few years, in her late forties, she has found herself feeling an overwhelming desire to learn about the man who raped her.

We compare notes. The gun. The mask. The command to stay silent. The presence of others. Still, there are many differences. In her case, she was on the second floor, and her mother and sisters and their friends were in the living room. She was alone in her room when she was raped.

We talk about our families' reactions. Although she was living with her mother at the time, her father was the one who took her to the hospital. The night after the rape, a group of kids from the neighborhood slept in her bedroom to protect her. Afterward, Lucy slept in her mother's bedroom for years. I feel an embarrassing shiver of envy upon hearing this. No mother held me close against the terrors of the night. Her parents sent her to therapy. They decided to tell the whole school what had happened, to prevent Lucy from feeling alone. She was frightened, she says, but she felt held.

Lucy has tried to get her file from the city of Cambridge, but she has made no headway. I tell her that I will look through my files, to see if there is any detail about the rapes that occurred in Cambridge. I have a vague recollection of a detailed list, I tell her. I will see what I can find.

A few days later, I make my way back to my garbage can. I leaf through the files quickly, pausing only long enough to determine whether the words "Cambridge" or "Radcliffe" appear anywhere on the sheets. In my mind's ear I hear the words, *Seek and ye shall find.* I push these words aside, embarrassed by the banality of my own mind. Other words replace them. *No pain, no gain.* Again, I brush the aphorism aside. As I leaf through the papers, I tell myself, I will not be distracted by the victims' pain. I repeat these words to myself, like a mantra.

At last I find what I'm looking for, a handwritten note that details the times and dates of the Cambridge rapes, and other rapes, taken down by Inspector Nestle, who was at that time investigating the rapes that took place at Concord Academy in 1971. He had discovered that there had been a series of rapes in nearby towns at about the same time, and he went to talk to the Cambridge police department to see if the rapes were similar.

The note is titled, "Info from Cambridge PD. Re: Rape Case." It is dated April 30, 1971.

Cambridge has had 20 incidents of rape and attempted rape, all with the same MO. These incidents began on April 9, 1970, at 9:35 PM and continue thru, to date. There was one long break in the pattern, from May to August of 1970. It was in late August that Provincetown had four (4) incidents with the same MO. Cambridge had it start up again on 9-2-70. The following are the dates and times of the Cambridge incidents for 1970:

4-9-70, 9:30 PM

9-2-70, 10 PM

9-15-70, 10:14 PM

10-10-70, 9:30 PM

10-13-70, 9:30 PM

10-28-70, 8:30 PM

10-28-70, 9 PM

11-23-70, 9:50 PM

11-24-70, 9:30 PM

12-2-70, 10:15 PM

In all of these incidents the subject's description and MO are the same. The subject's description is as follows: White male—5'11" to 6'2". His weight runs from 135 lbs to 155 lbs. In all incidents the eyes are described as "clear blue."

Subject uses a gun. It is said to be a small black or dark grey revolver with white grips. After having the victims disrobe, and lay down, he uses a salve or lubricant from a tube silver or grey in color (there is a possibility this is medication for the treatment of gonorrhea—several of the Cambridge victims have come down with a strong strain of it) on the genitals, and then has intercourse with them. All victims state that the subject appears to be fairly well educated, well spoken, with a medium pitched voice. He is very methodical in his movements. The victims all state that the subject is very nervous, almost to the point of being as scared as the victims, and he always is polite and apologizes to the victims for what he is doing and he keeps stating he doesn't want to hurt them.

The following are the Cambridge incidents for 1971:

1-8-71, 9:30 PM
3-15-71, 8:30 PM
3-16-71, 9:20 PM
3-16-71, 9:55 PM
4-2-71, 9 PM
4-12-71, 9:45 PM
4-21-71, 9:50 PM
4-22-71, 12:05 AM

Same MO as the description of the 1970 incidents.

While at Cambridge—read the reports from Provincetown PD on the 3 incidents they had in late August. Incident involved a subject fitting our subject and the Cambridge subject. Girls involved were young teenagers. In an attempted rape in Canton, spoke with Sgt Reno about incident in that town. W/M 5'11" 140 slender build. Blue eyes, mask, mask taken off. Same MO as our incident. Canton incident was on 3-12-71 at about 9 pm.

In going over the incidents (Cambridge and Canton and Provincetown) they are very similar. The only missing items are the gloves and rocks. These two items are in our incident [the rape at Concord Academy that Inspector Nestle was investigating at the time]. The rocks show up in the Natick incident, which occurred on 4-26-71. The other similar items are the use of salve, the gun, and the masks. All the victims are aged 9 to 19. The Cambridge incidents involved girls who live in an eight-block area around Rad-

cliffe college, some of the incidents were in the dorms, the others in houses used by students living in the area. The Natick incident was at a private girls' school, the same as our incident. The Provincetown and Canton incidents were in private homes.

In going over all the available info, there is no question that all these incidents are almost identical, the only marked difference being the rocks used in Natick and Concord, and the gloves used in Concord.

I send the list of dates to Lucy. She was raped on March 15, 1971, at 8:30 PM. Now she tells me about the salve, which she was too embarrassed to tell anyone before, even her mother. There was something particularly disgusting about it. And she tells me about the rapist's piercing blue eyes, which she says she will never forget.

Lucy's reaction, and the reaction of her family, is surprising to me. Her sisters treat me like a hero, rather than the troublemaker I know myself to be. They send me thank-you notes. They are interested in what I am writing, and when I send them parts of my manuscript, they respond almost immediately. We get together for brunch, as if we were gathering for a family reunion. Her sisters tell me that because I've helped Lucy find the identity of her rapist and because we've discovered that the rapist is dead, Lucy will live with less fear.

All this softness, at a time like this, is almost hard for me to take. I feel held, even loved. But I am afraid to express to Lucy and her sisters how much their reaction means to me.

I visit Lucy at her home. She lives on a cheerfully busy street in Burlington, Vermont, in a big, rambling Victorian, reminiscent of the rambling Victorians you find in the part of Cambridge where she grew up. We are sitting in Lucy's living room, our stocking feet resting on a chair in front of her woodstove.

Now that we're pretty sure we were raped by the same man, Lucy and I have more detailed conversations.

"Why did you get curious about your rapist?" I ask her.

"I was coming up on my fiftieth birthday. I guess that was important because that was the age my mother was when she died. And then my father died. When we were going through his papers, I found a subpoena of one of the suspects. On the back of the subpoena I found a diagram in his writing. A diagram that showed where my room was in the house.

"So I found this subpoena," she repeats.

She switches back to what was going on in her life when she first became curious about her rapist.

"My marriage had started to fall apart. I felt betrayed. Three years ago. And that brought back the memory of the earlier traumas, my rape, my mother's death. For months I had a feeling, that terror feeling, that feeling in your stomach."

"What do you mean?" I ask. I never thought of any of the feelings I get when I'm afraid as a "terror feeling," and I want to know what Lucy feels.

"It's like a spark of fear, the feeling you get when you see the police pulling up behind you. But it didn't go away, it stayed in my stomach."

With a gesture of her hand in the air, she indicates a line in the air from her lower abdomen to her heart.

"It hurt so much," she says. "I couldn't relax my abdomen."

I imagine Lucy's muscles clenched tight, her abdomen trying to muscle something out, but actually forcing it deeper. I am listening to Lucy, but in another part of my mind, I carefully remove the pain from her abdomen.

"My marriage had fallen apart, and there was no reality in my world anymore. I was afraid. It brought back an earlier period when I was afraid all the time. My father and mother split up when I was in seventh grade. My father remarried in June of my seventh-grade year, and in March of my eighth-grade year I was raped."

"I think he was being a Peeping Tom before he raped me," she adds, switching back to the rapist.

I don't ask her what she means by that. Maybe I'm too frightened. Did he case her house before he raped her, the way he scoped out ours? Before my sister and I were raped, the rapist had figured out how to cut off the telephone. In one of the files I read through quickly, a woman on the Cape came home to find Brian Beat in her home, apparently casing her house in preparation for a future crime. She saw him clearly, and was able to identify the intruder for the police. He would later be convicted of raping a woman who lived nearby, as well as breaking in and entering her home.

"You might have thought it was a dorm where we were living," she says. "It was 1971. Cambridge. A big old Cambridge Victorian. There were eleven women in the house at the time I was raped.

"It was like there was a misfiring going on in my brain. All of these things were connected," she says. She has switched back to her life of several years ago.

"My marriage was falling apart. I felt the same way I did after I was raped, the same way I did after my mother died. I lost a ton of weight. I couldn't get my stomach to relax. As I sit here today I'm so happy, I'm so happy, I'm so happy I don't have that feeling in my stomach. . . ."

At first, hearing these words, I am puzzled. Why is she so happy? Then I realize she is talking about the pain. The pain had stopped.

"As I got into trying to understand this feeling, I started to realize that I was very afraid. I was dealing with something that I was afraid of—I could meet the guy that raped me, I could . . ."

Again, I'm confused. Is she talking about the feeling in her stomach? I sip some of the coffee Lucy has poured for me, needing the caffeine, hoping it won't give me a stomachache. I have not been sufficiently vigilant about the ways her story might alter my own perceptions, might make it hard for me to stay in the room.

"So my dad leaves. My mother falls apart. I start eighth grade, and then in March I get raped. My mother dies. And now, many years later, my marriage was falling apart. It brought back all these earlier wounds," she says, the many causes of the pain in her abdomen all jumbling out at once.

"Ever since the rape I have felt afraid to be alone in a big house like this. I slept in my mother's room for years. If I was home alone at night, I would hear things. Dark big house, people in the big house . . ." She tells me that the place she felt safest was a studio apartment in New York City. "It was tiny," she says. "On the sixteenth floor. No place to hide."

Lucy has not told her two teenage boys that she had been raped. We talk about the pros and cons. I wonder whether children sense something, but don't know what they sense; whether they might imagine that they are somehow to blame for their mother's fear.

"Do you think the rape might have changed the way you were as a mother?" I ask. Of course I want to know the answer to this question for myself: How might it affect how I parent my son?

"I don't worry that they're going to be raped," she says, answering a question I haven't asked. "They are huge strapping boys. But I worry that something bad will happen to them. Of course it's unlikely, but so was what happened to me.

"How likely was it in 1971 that a thirteen-year-old girl would get raped at gunpoint with her mother and nine other women in the house? Even today, in my old neighborhood, in Cambridge, this sort of thing doesn't happen. It was something we couldn't even imagine back then.

"I am overprotective," she adds. She tells me about a chess coach her boys had; that she sensed that he was a predator. "I didn't get a good feeling around him," she says. "I didn't want the boys to be with him."

"How did your husband react to that?" I ask.

"He thought I was being overly paranoid," she says. "He was dismissive.

"But I never let my kids be with the chess coach by themselves. They would say, 'Mom, he's not a pedophile.' It turned out that he was. He's now in federal prison."

She may have been paranoid, but her intuition was correct. Does it take a victim to recognize a predator before anyone else does?

"How did your husband react when you told him you were raped?" I ask.

"He was compassionate. I think he felt bad about it, but it wasn't real to him.

"For many years, I felt ashamed of myself. Ashamed of my own body."

I see her looking out the window, trying to look away from an uncomfortable feeling.

I want to protect Lucy. She looks so slim and vulnerable. And today, while we are talking, I see a little girl.

But at the same time, in this moment, I will confess that I feel superior to Lucy. Emotionally healthier. More grown-up.

I am not aware of feeling ashamed of my own body in the way she describes. But I am aware of feeling ashamed of this feeling I have now, this slightly superior, competitive feeling with Lucy.

Now that I hear about the possibility of harboring this kind of shame, I imagine that I must have it, too. Like the medical students who imagine themselves with laryngitis, leukemia, lymphoma, every disease in the medical textbook. Perhaps I have repressed the feeling of shame over my own body. Perhaps I have muffled it. Or "forgotten" the many times I've been aware of feeling it.

But now, reading over my notes, I realize that this need I have to clean things—not so much when they are dirty as when I'm in the mood—might be connected with shame. This morning, for example, sweeping the hearth, I was determined to get the ashes from the spaces between the bricks. After I swept the hearth, I swept the leaves off the porch. Then I wiped off the kitchen counters, which were clean, with orange-scented soap.

"Why is rape so shaming?" I ask her. I am certain that rape is shaming, though I'm less aware of feeling ashamed of my body than Lucy is.

But now, again reading over my notes, I think back to the period before the rape. I was dancing a lot back then. Ballet and modern. The feeling of soaring. So much music in my life. After the rape, I remember the shift in my posture. Somehow I noticed it, as if I could stand outside myself and observe. The shoulders slumping forward, as if I had something to hide. How long did it take for me to shift from dancer's posture to the posture of a victim?

And now that I've uttered that word, I am truly ashamed. I did not have the posture of a victim. I was not a victim. I was just raped.

"People say that rape is not sex, that it's violence," Lucy says, bitterly. "But it's also sex. You can't get around that," she says. "He didn't run me over with a car. He had sex with me. You're not supposed to do that. You're not supposed to have sex with an eighth-grader. You're not supposed to have sex when you're in eighth grade. It was very intimate. You can't get around it. This part of the body," she says, gesturing from her heart to her lower abdomen, though I understand she means to indicate her vagina. "If you're sitting around with a group of women, talking about various traumas, someone will say, I got beaten by my mother. But if you say, I got raped, it's a different thing."

I wonder if that is true. Is rape really the worst sort of violation? I'm not sure. I often wonder why it matters whether we're penetrated or not. There is the pain, but the pain doesn't last. The shame does.

And I realize that for me, rape didn't seem like sex. It seemed like a discharge of shame, an exchange of pain. Rape changes ones sexuality. Of course it does. It must. I'm terrorized when men are attracted to me. But I cannot write about my sexuality. I feel too vulnerable. I am afraid that if I call attention to myself in that way, I might be raped again.[4]

Now I want to ask her about her memory of the rape itself.

"What do you remember about the gun?" I ask, wondering if she remembers a white handle.

"I remember a short dark gun," she says. I sense her entering the memory now. My body tenses. I will myself to stay alert.

"It was March fifteenth. I remember because it was the ides of March. It was unseasonably warm. The windows were open. It felt like an early summer day. It was too hot."

I ask Lucy if she dislikes this kind of heat now, when it's unseasonably hot. "I love the heat," she says. But I don't believe her.

"I remember that I walked home with my friends from Harvard Square. We all had dinner. Or maybe we had dinner in the square, I'm not sure. I know there were eleven of us in the house at the time. There were my two sisters. Jen had two friends over. There was my mother's friend Polly. My friends. That makes nine. We may have had two other girls living in the house. We often had kids staying with us. So there were all these women in the house. I went upstairs. . . . I was raped at eight fifteen," she says.

She pauses. I can see that she is trying to get this right.

"That's right," she says, a faraway look in her eye. "It was our spring vacation. There was no school then. So I came up to my room. I don't remember where I got undressed. I went into my mother's bathroom. I took a shower in her bathroom. I must have wanted to use her stuff, or maybe because it was a better shower.

"I turn the shower off and step out," she says. She has switched to present tense. "As soon as I step out, I see a man wearing a dark blue slicker and a black mask with cutout eyes. It looked like a ski mask. He had very blue eyes."

We are back in the past.

"I thought he was a friend of my sister's, making a joke. I reacted that way at first. He had a gun. I said to him, 'What are you doing? Get out,' still thinking it was a joke. He didn't respond accordingly. That was the moment that I realized that he was a man that was nobody I knew."

Why did the police have such a hard time believing my sister and me that the man who raped us was "nobody we knew"?

And now she says, "I remember that feeling coming over me—"

Lucy looks out the window. She is bothered by a noise, which I haven't heard.

"This house is too big for me," she says. "They are chopping down the trees outside."

She runs outside for a moment. Men from the telephone company, she says.

She returns, back to the too-big house. I see a faraway look in her eye now. It seems to me that her features are slightly altered.

"Is this really okay?" I ask.

She assures me that she's fine. She returns to her story.

"At the moment when I realized that he was nobody I knew . . ."

She pauses. I hear the ticking of the clock. She finds her way back to her story.

"There is the moment when the blanket went over me, of disconnection with my body. Like I was almost floating at that point. I don't remember exactly what he said, other than telling me, Don't say anything. He told me to go through my mother's room. To go out the door of her room and through the door into mine. There was a lock on the door from years before. A bolt lock. All I had on was a towel. He told me to lie down on my bed. My room had two beds in it. I don't know how he knew which was my bed. I can't remember what he did with the gun. . . ." She pauses.

I wonder to myself, Is she considering now whether she might have been able to grab the gun, might have been able to kill him? Whether she might have tried to run away when he put the gun down? I don't ask these questions. I am afraid to interrupt her narrative. I do not believe I would be able to remember as much as Lucy has if I didn't have the material I wrote down for the police at the time. I would probably not know what happened to me. I would have forgotten.

"I remember he got something out of his pocket. It was a tube of junk," she says. What a harsh word, I think, the hardness of that final *k*. I feel the pain and the shame of the cold ointment on her vagina, the vagina of a thirteen-year-old girl.

I can see that she cannot describe what the "junk" was. Somehow it is this "junk" that horrifies her more than anything.

"He put it on me and on him," she says. She looks far away. Her hand floats up to cover her mouth.

She recovers herself. "I said to him, Are you going to rape me?"

I notice Lucy's long, lean legs, stretched out onto the stool in front of us. Her anger gives her strength, I think to myself. She will live through this.

"And then I don't remember what he said."

She pauses again. There is some thought here that is too painful to capture out loud.

"I do remember feeling his penis," she says.

I am shocked awake by that word *penis*. It sounds so technical, so grown-up. It seems so out of keeping with the way Lucy has been speaking until this point.

"It was very, very painful. Like he was forcing something that is too big into a too-small hole," she says.

I do not like this word *hole*, referring to a child's vagina. I want to edit her recollections for her, but I don't.

"I don't remember the act itself," she says. And then adds quickly, "Then he got up and I think he just went out."

I see her switch again. She is finished with something; she is ready to be done with this topic.

I ask her whether she remembers having to put other clothing on, whether she remembers any stones.

"He didn't make me do anything other than what I just said," she says. "He didn't spend a lot of time; there were other people in the house." Perhaps he couldn't take the risk that she would scream. There was no one there to muffle her mouth.

"I remember that I lay there for a little bit. On my bed. Maybe ten seconds or a minute. I opened my door. I remember the feeling—a kind of cloud of haze. But when I got to the bottom of the stairs, I started to scream."

She continues, "I started screaming, 'I got raped.' At first they thought I got cut in the shower.

"After that I don't remember very much." I see a quiver of pain in her eye, but no tears.

"I do remember one awful thing, though," she says. "The bedpan. I remember getting into a cruiser and going down to the hospital. We went to MGH. I don't know why we didn't go to Mt. Auburn, the hospital nearby. They examined me. They pulled a curtain around me. I remember thinking that other people might be able to see me, that I wasn't well protected. They put a hard bedpan under my bottom, to raise up my body, so they could see. It's hard to believe that they would do that to a little girl. A cold, hard bedpan, under me. Why did they do that?

"I have my ER record. We had to get prescriptions. I'm not sure what for. I know I had to take the morning-after pill. I threw up all of the next day.

"My father took me to the hospital." She hands me the report.

When I gave Lucy the list of dates and times of the rapes that had occurred in her neighborhood in 1971, she was furious. If only the police had informed the neighborhood, she said, she probably would not have been raped. There had been eleven incidents in an eight-block area in the neighborhood of Radcliffe, prior to Lucy's rape.

"You would have thought that the police would want everyone to know that there had been a series of rapes in the neighborhood, to make sure everyone knew, to protect the other girls," she says, angry again.

I wonder, Do I sound angry about my rape? Do I still come across as matter-of-fact?

"But when we asked the chief of police at Harvard," she says, referring to a conversation she had had with him after she and I started communicating, "he said that it was just a different time. People saw rape differently back then. And for the police—it was soon after the protests. They had different priorities.

"I think he also meant that the officers were not as equipped psychologically to deal with rape as they are now," Lucy adds. We are back in the present.

"When I saw the photograph, I knew that it was my rapist. It was his eyes and his pockmarked neck. His eyes were very bright."

A part of me is oddly envious of Lucy in this moment. The fact that she can remember the rapist's eyes. I remember so little.

I haven't looked very closely at the pictures of the rapist. The first time Lt. Macone warned me there were photographs in the file, I turned the page instantly. I did not want to look. And even now, I rush through the files when I know I'm near a photograph.

Recently I found a photograph of him at the time he was arrested, and I realized that the photos I'd been seeing were from the time he was in prison, when he was older than he was at the time we were raped. This new photograph is of a thin man, with light skin, a long face. He looks vaguely familiar. I feel a chill, so I don't look more than a fraction of a second. I might almost say that I recognize him. But is he familiar from a dream? Is he familiar from what I remember, or don't remember, from 1973; or from a photograph that I glimpsed in the file, in 2008? I will probably never know. At this point I cannot distinguish my memory of the rape from my recollections of seeing the images and reports in that file. Eventually I will work up the courage to ask my sister, but I don't want to ask her now. I don't want to hurt her.

"When we start talking about penetration—" she says, but then stops.

She begins again. "Penetration. It creates this horrible shame. . . . It's a barrier of your body, a barrier that gets broken," she says.

Once again, I notice that her eyes are slightly swollen. I always have the feeling that my own eyes swell when I'm thinking about my rape. And now I wonder, seeing Lucy's face, whether the swelling I imagine is actually real, even if it's too subtle for most people to notice.

We return to the topic of terror.

"That physical feeling. In my abdomen. The first time I had the feeling was right after I was raped. And then again, when my mother died."

"My mother was cut open from here to here," she says, indicating that same part of her body she pointed to earlier, from her lower abdomen to her heart. What is she talking about? I never heard that her mother was murdered.

"Her pancreas had ruptured. In the final attempt to save her, they had to cut her open. They left her open to kind of drain."

I am horrified by this image, but relieved that it was doctors who had cut her mother open, rather than a rapist.

"I wasn't supposed to see my mother like that in the hospital. She was covered up, but of course we could still see that she was cut open. She was covered in a kind of nightie, but we could see through it. When she moved. Seeing her like that.

"That is what the feeling is like," she says, "like you're cut open. First, it's the feeling of being startled. All your nerve endings, they jangle. Right after that there is that other feeling, an electric feeling. You know what I mean, your stomach is jazzed up. It's a secondary reaction to the fear. It goes from your lower belly all the way into your chest."

The thought of this feeling brings her back to the rape, the subject that had been temporarily closed.

"I remember barfing all day long the next day, the day after the rape. Maybe in my mind I put all this together, my mother's belly sliced open, the rape, the barfing. And I got that same pain when my marriage began falling apart."

Now I know why the word *penis* sounded so alien to me. It is because Lucy is using the language we used as teenagers—"barfing," "your stomach jazzed up," "junk" to refer to something too "yucky" to name. It takes me back to my own time of terror, uncertainty, abandonment.

Lucy continues. "If it came up in conversation, I couldn't say the word *rape*," she says. "It got stuck in my throat. I just couldn't say it. If I said it, if I had to say it, I felt like I was choking. I prefer the words *sexual assault*, or *aggravated sexual assault*.

"There is something very primal about rape, like cavemen. It's a yucky word."

"What did you think when your stepmother told you that she had met a woman who had been raped around the same time you were?" I ask.

"She e-mailed," Lucy recalls. "It didn't even occur to me it was the same guy. But I remember thinking it would be nice to talk to someone who had been through the same thing. I was visiting my cousins. They were asking me, Why would you want to meet this person, a random person who was raped? And I said, You have to understand that there is one person whom I know in the entire universe who had the same experience. It is a huge universe, and I'd never met anyone who was raped like I was, at gunpoint.

"It felt so strange that it had come to me now—this possibility of meeting you—because it had been on my mind, and I was frustrated that I had to let it go. . . . I had to let it go because I had hit a dead end in my search for the files."

The Cambridge police cannot find the file. But Harvard University has its own police force and keeps its own records. So Lucy and I requested Harvard's records. We wrote to Harvard's president, who sent our request to the counsel's office. It took the counsel a year to respond to our numerous calls. He was busy, he explained. Perhaps he was too busy to redact the files. But would he have been too busy to help us if he knew who Lucy is, who her father was?

"I think it's my father's divine intervention that we found each other," she adds, unexpectedly. I would not have guessed that Lucy believed in divine intervention.

"My father has a hand in things," she says.

What I learned about my rapist:

After reopening a thirty-three-year-old cold case, the police are now persuaded that they have identified the man who raped my sister and me. Brian Beat was convicted of three earlier rapes and was sent to prison for eighteen years, only ten days after he raped us. Although the police knew of this man and had in their files information on a remarkably similar series of crimes that had occurred in the Boston area in the two years prior to our rape, they did not believe us when we insisted that we did not know the man who raped us. They did not put the pieces together, and for thirty-three years, the crimes remained unsolved.

Brian Beat committed suicide several years before I began looking for him, so I was unable to interview him, and he was unable to confirm or deny our suspicions. But I was able to talk to many people who knew him, and to get access to many of the prison, court, and police records associated with his crimes. Based on the similarity of the description of the perpetrator, his weapon, and his unusual modus operandi, the police are now convinced that Brian Beat raped at least forty-four girls in Massachusetts between 1971 and 1973. The girls were between the ages of nine and nineteen. Most of the rapes were at or near private girls' schools. Many of the rapes involved two or more girls.

People who knew Brian Beat in his youth described him as a "gorgeous" and "brilliant" young man who was "nice" most of the time. But he was also capable of sudden, unprovoked acts of cruelty that seemed to come from nowhere, as if he had become another person. Although most of his relatives and friends focused, in their conversations with me, on the side of him that was "nice," I learned that girls who were part of his circle were warned not to be alone with him, and that before he began his career of raping unknown girls at gunpoint, he tried to rape the sister of one of his best friends.

All but two of Brian Beat's friends and relatives insisted, at the beginning of our conversation, that he was innocent of any crime. But almost invariably, they would state, as if as an afterthought, as if only vaguely aware of the significance of their words, that he should not have done what he did (without specifying what that was), or that they knew he was "disturbed."

I cannot know what combination of biochemical or psychological or other factors caused Brian Beat to do such monstrous things. But I was able to learn this: He left a wake of suffering among many of those exposed to him, and he was himself traumatized. He was adopted by an aunt and learned of his provenance in the worst possible way, from a child on the playground whose aim was to hurt him. As a young teenager he learned that the woman he thought was his aunt was actually his birth mother; and that his cousin was actually his half sister. Although he was not openly homosexual, at least among the group of his friends that I was able to meet, he frequented gay bars and was excused from serving in the military because he claimed to be homosexual. He lived in a part of Massachusetts that was a dumping ground for pedophile priests, where members of the clergy passed their victims from one pedophile priest to the next. He attended a church in Milbridge, Massachusetts, that suffered a series of predatory priests. These abuses were brought to light after the church scandal broke in Boston in 2002. There were rumors about sexual abuse at his elementary school at the time he was a student there, but no case was brought to court at the time. The House of Affirmation, located in a nearby town and also part of the Worcester Diocese, was founded in 1973 as a treatment center for priests with psychological and sexual problems. However, a number of the priests residing there for treatment continued to engage in the sexual abuse of minors from the neighboring towns and became the subjects of multiple criminal investigations and civil complaints and settlements.

Clergy sexual abuse was rarely discussed in the 1950s, and I was not able to identify a single court case from the period of

Brian Beat's childhood. It is not possible to know whether the alleged pedophile ring associated with the House of Affirmation was active prior to that organization's formal establishment, or if pedophiles were routinely placed at Brian Beat's church in the 1950s as they were in later years. However, the rituals associated with some of Beat's rapes—including the stones he placed at the scene of the crime and the small stones that he routinely carried in his front pocket—suggested that he was reenacting a ritual of some kind.

After serving eighteen years for three of the forty-four rapes, Beat returned to the small town in central Massachusetts where he grew up. The police did not get around to designating him a sexual offender under Megan's Law until he began harassing a local police officer. His case fell through the cracks.

chapter twelve

War's Victims

My shrink tells me I have PTSD. Ridiculous. PTSD is what soldiers get, not girls.

I know there are certain things that set my teeth on edge. Fluorescent lights. Greasy fingers. That ticking, scraping sound that overwhelms you with rage. When I'm in my analyst's office, the sound of her clock intrudes into my skull. I cannot bear the sound.

But still. The impact of war and the impact of sexual violence are not the same.

Certain smells make me depressed. The smell of air freshener can make me want to die. It's also true that I am spacy sometimes. Can't drive. If I get lost, it feels like the world is caving in on me. Can't think sometimes. My mind. It drips somehow. And I will admit I've done some dangerous things. But with good reason. My research.

Is it really possible, as experts say, that rape or long-term "relational abuse," to use a shrinky term, can have a physiological effect on the mind and body similar to that of the terror of war?

Researchers armed with magnetic resonance imaging devices repeatedly report that they can see and measure physical changes in the brain associated with PTSD. They can measure significant reductions in the volume of the hippocampus and changes in the medial prefrontal cortex, areas of the brain that are involved in memory as well as emotional response. These physical changes have been observed in Vietnam vets and in victims of childhood physical or sexual abuse. There was no change in vets or victims of abuse who did not suffer PTSD. Reduction in hippocampal volume is specific to PTSD, and is not observed in other panic or anxiety disorders.[5] I have read this again and again. But I can't see my hippocampus. And it seems to me that PTSD is a fashionable disorder.

I'm not going to take the word of a shrink. I'm going to talk to some soldiers and see for myself.

I have a friend whose work puts him in regular contact with soldiers. Karl is a retired Special Services officer whose company does a lot of work in Iraq, and he knows a lot of vets. I ask him about the possibility of talking to an Iraq War veteran with PTSD. Sure, he says.

Karl e-mails me the contact information for his son-in-law, who has complained of PTSD symptoms since his return from Iraq. Karl had mentioned his son-in-law Erik earlier, when Karl and I first met. We had earlier talked about my traveling down to Fort Bragg to one of the hospitals there. But his son-in-law lives relatively close by.

I think I better tell you the backstory of how, when, and why I first met Karl. Karl retired after serving for twenty years in the Special Forces. After that, he started his own company, a military contractor. The war in Iraq was good for Karl's business.

He won a significant contract with the Department of Defense to help oversee American detention facilities. After General Petraeus's "surge," the American military found itself with some 26,000 Iraqis under detention, and Karl's company became involved in overseeing their confinement.

The large number of detainees was causing problems for both the U.S. and the Iraqi governments. Some 85 percent of the detainees were Sunni. The Sunni bloc in the parliament was regularly threatening to walk out. General Petraeus was anxious to find a way to "rehabilitate" the detainees and get them back out on the street as soon as possible. He hired Karl to develop a plan for rehabilitating the terrorists and insurgents, and Karl proposed to hire me. I was skeptical about the possibility that Americans would be able to rehabilitate professional terrorists, but I was happy to learn about the work.

"I can see, Dr. Stern, from reading your books, that you believe in acquiring ground truths," he begins.

Karl refers to me as "ma'am" or "Dr. Stern," even though I insist that he call me Jessica. He will comply for five minutes, but he always reverts to a more respectful form of address. I find this distracting, but also touching. Karl cannot give up the habits of a soldier.

The phrase *ground truths* has a heavy sound to my ear. But the concept is appealing to me. I don't like to rely solely on information contained in libraries or other people's field research, which is what I was trained to do. I want to roll up my sleeves, immerse myself in the mess of the moment, allow the truth to reveal itself in the stories that inevitably unfold. How does he know this about me?

And then, there is that word *truths*. How does he know how to flatter me so well? He sees that I am obsessed with finding truths, and not only that, that I understand that there is no one "truth." He has identified precisely the qualities I am most proud of in my work. I am naturally skeptical of established facts. I am fascinated by the details of people's lives. I like to bathe in the data, to look at every detail, to discover the secrets that a more single-minded, disciplined researcher would surely miss. Karl is skilled. He does not flatter me in an obvious way, in a way that would make me feel manipulated.

What he needs, Karl tells me, is a person who knows how to talk to terrorists. "A person capable of getting beyond the political jargon," he says. A person who is capable of discovering terrorists' actual motivations, not the rationale they might publish on a Web site.

I am skeptical that professional terrorists can be rehabilitated by outsiders, I tell him again. But I know that a large fraction of the detainees are insurgents, not terrorists, in the sense that they are fighting an occupying force, not shooting civilians.

The idea of learning about these detainees is extremely appealing to me. To put it bluntly, I am fascinated by "bad guys." I imagine myself reading over interview notes, the results of questionnaires. Perhaps I might even be able to write the questionnaires.

But now comes the most important part. The work requires spending time in Iraq.

"You would get to choose where you work," he tells me, "at Camp Cropper, near the airport, or Camp Bucca, near the border with Kuwait."

I immediately protest. "I have a young child," I remind him. Karl has already seen the evidence of my son in the form of toys strewn about my home office. "I cannot expose myself to this kind of risk anymore," I tell him. I even confess, defensively, that because my mother died when I was three years old, I am especially aware of the impact of a mother's death on a young child's life, and perhaps unusually wary of exposing myself to risk.

I want very much to say yes. I am seduced. First, there is the flattering thought that I am uniquely qualified for the job. Next, there is the gratifying thought that I might be able to save lives. Then there is my dangerous curiosity, which I cannot seem to shed. Finally—and I am not proud of this—there is the appeal of danger itself. I know what would happen to me if I were truly endangered. I would not actually feel my fear. Instead, I would be excited, smart, efficient. It is like a drug, this feeling that I get when I'm exposed to the threat of death. I would be capable of doing my very best work, at least for a while, until exhaustion set in; or until, if I were unlucky, I was killed.

I can feel, though I cannot tell you how, that Karl is very much like me, at least in this one way. I feel a rush in his presence. It is as if we are in sync—like two waterfalls that find a way to flow at the same speed. I am searching for the right words, and I'm afraid you might imagine that I'm speaking, rather confusedly, about sexual attraction. I am speaking about chemical attraction, that is true; but it is much broader, and more diffuse, than lust for a person. As he is talking, I can *feel* Karl's excitement about this job. I sense it in my own racing heart. I don't know what it is about certain soldiers that evokes this feeling in me, but I'll take a stab at identifying what it is. Karl is excited by exposing himself to danger, just as I am. Not for a foolish cause, but to make the world a better place. When I'm around a person like Karl, a person who does good works in dangerous places, I feel the urge to join the cause. The question is, Are we capable of distinguishing which dangerous jobs will in fact improve the world, and which will not?

I have already said no, and I mean it. But Karl continues with his seduction.

"I keep my staff alive by educating them," he tells me. He

is proud that not a single one of his personnel has been killed.

And then he makes the most astonishing offer. "I will provide you counterterrorism training," he says. "At a facility near Fort Bragg."

"You mean at a Special Forces facility?" I ask.

"No, just nearby," he says.

I am embarrassed to report all this. In this moment, I am like the sweetheart in Tim O'Brien's story "Sweetheart of the Song Tra Bong."[6] I am fascinated and seduced by the apparatus of war, especially the process of acquiring and analyzing secret information. I spent my postdoctoral years working as an analyst in a nuclear weapons laboratory.

I love the idea of learning how to trump terrorists at their own game. My violent fantasies run in the direction of evasion and trickery, not weaponry, but I suppose that is appropriate, given the stereotypes attached to my gender. In my mind, Karl is proposing to turn me into a modern Mata Hari, not a security guard. He wants me for my brains, not my brawn. Maybe they will teach me how to disappear, I think to myself, or to hypnotize my prey. I try to discipline my unruly mind, to be serious, to exude gravitas. How ironic that I am being offered this training now, rather than before, when I was meeting with terrorists one-on-one in Pakistan and Lebanon.

Can you imagine how seductive this offer is to a person who studies terrorism? To get this kind of training? Can you imagine what this means to the daughter of a refugee from Nazi Germany? To a victim of rape?

Briefly I consider asking Karl whether he would allow me to receive this training even if I didn't go to Iraq. But I put that thought out of my mind. Too selfish. A waste of valuable resources. And anyway, I tell myself, I couldn't do it. Too many days away from my son.

As if reading my mind, he reassures me, "Part of the training can be done one-on-one, so you'd be able to remain in Cambridge with your son.

"I will send the trainers to you," he repeats, as if we have closed the deal.

For a moment it occurs to me that I am facing a kind of devil. Here is how Temptation appears to me: A Special Job that makes the world a Safer Place, but that is dangerous for me. Only those with special training (and smarts) can survive. What it takes to excel is hypervigilance and intuition. The ability to tame killers and men who carry guns. The ability to remain extremely calm under threat.

"We'll fly you commercial into Kuwait," he continues, "and then by military transport into Iraq." I am embarrassed to admit that even the idea of "flying military transport" appeals to me.

"There is very little risk," he assures me. He also tells me, "My whole family has been to Iraq. My daughter has been there. My son-in-law is there now."

Attraction to danger is a common response after exposure to life-threatening trauma. I want to argue here that this characteristic, if channeled appropriately, can result in extraordinary public service; that not all the "symptoms" associated with exposure to trauma are bad for society or even for individual sufferers.

In the end, I did not go to Iraq. I took a desk job, helping Karl part-time from Cambridge. But if this offer had come to me at another point in my life, I would have jumped at the opportunity. So today Karl's son-in-law is back to haunt me. What should I feel?

Erik and I have a few preliminary conversations, and eventually we agree to meet in a town located between where he lives in Maine and where I live in Massachusetts. We agree to meet at the home of one of his friends, a fellow soldier, another Iraq War veteran. I take Route 93 up north. That's easy. Then I drive just a few miles off the highway. There are only four additional turns. I make it without mishap.

When I arrive at Erik's friend's home, I dial Erik's cell phone. He ambles out of the house, looking as if he just woke up. He is wearing one of those sleek jackets, warmth without bulk, the kind you get at Patagonia. The green of the jacket sets off his olive skin. He is good-looking, I observe, in a kind of wholesome way, despite an apparent hangover. I notice hiking boots on his feet. He must have worn combat boots in Iraq. I wonder if he likes to hike. It is 11:10 AM.

He suggests that we go to a coffee shop near the highway. Dunkin' Donuts again. All my interviews seem to take place in Dunkin' Donuts.

He gets into my car, directs me back toward the highway. As we are driving, I try to make small talk.

Erik is having trouble making ends meet. He has two jobs. I know that he drove down to his friend's house in New Hampshire last night, that he has a rare weekend day off.

"What did you do last night?" I ask, trying to feel and sound bemused by his apparently sleepy state.

"We were playing drinking games," he says.

"What are drinking games?" I ask, feeling awkward and confused.

He tells me that he and his buddy were playing cards, telling jokes, and drinking until morning.

I tell him how grateful I am that he is willing to talk to me, how grateful I am to his father-in-law for introducing us.

"He's not really my father-in-law," Erik says. "Not anymore. Betsy and I are divorced."

I wonder why Karl didn't tell me this.

"The last time I came back from Iraq, I found out she had been doing bad things. But I always liked Karl. He's a great guy. A brilliant guy. I have no problems with him, only with his daughter."

I don't ask what "bad things" means. I know that a high percentage of Iraq vets' marriages end in divorce.

When we arrive at Dunkin' Donuts, Erik orders a large coffee, a large cup of juice, and an egg sandwich with bacon. He seems famished. I wonder if he eats enough. He wants to pay for his own meal, even though I am the one who has torn him away from his normal life.

He has chosen a table in the back of the coffee shop.

"Where did you grow up?" I ask.

"In Maine," he tells me, close to where he lives now. "In Union."

"How did you end up in Iraq?" I ask.

He tells me he wanted to be a chef. He applied and was accepted to Johnson and Wales, which I will later learn is considered the Harvard University of cooking schools.

"At that time, the army was running an experimental program called College First," he says. "They paid us a hundred and fifty dollars a month, and promised to pay back our federal loans if we signed up with the army when we finished our degrees. Four years," he says. "We had to join up for four years."

"What if you didn't want to join when you finished your cooking degree?" I ask.

"We'd have to pay the stipend back," he says. "Everything they gave us, plus interest."

Erik's dream was to open his own restaurant in Camden. Camden is perhaps the fanciest resort town in Maine, popular with wealthy New Yorkers with refined palates. If you have a cooking degree and want to be a chef, Camden would be a good place to start out. But when Erik finished his degree, there were no cooking jobs to be had in Camden.

"The army told me they would pay off all my loans if I joined the service, and I didn't see a way to pay them back myself. They told me I'd be skydiving," he said. "Jumping out of planes from ten thousand feet. All this cool stuff. But I never did any of that. They lied to me."

He has some trouble telling the story in chronological order.

"I was injured by an IED [improvised explosive device] in Iraq," he blurts out.

This is the first I have heard of any physical injury. Karl did not tell me that his daughter divorced the man who was his son-in-law. He did not tell me that Erik was injured.

"Right outside of Tikrit. Blew out my right eardrum. I had skin grafts. My arm was completely mangled," he says.

Without my asking, he rolls up the sleeve on his right arm. Two large sections of his arm are red and raw, as if, in remaking Erik's arm, the surgeon had run out of human flesh and was forced to substitute a side of beef. Instinctively I pull away, and then quickly lean in, embarrassed by the cruelty of my animal reaction to his wound.

Now he points to his face. I note his blue eyes, long lashes. A sweet boy, I think to myself.

"All this is dirt and shrapnel still stuck in my face," he says, pointing to flecks of gray and black metal embedded in his skin. Still, his eyes make the strongest impression. Vulnerable. Also angry. I don't know why it should occur to me, but maybe he's not angry enough.

"My right mandible—"

He starts again. "The whole side of my right jaw was completely shattered. They wired my mouth shut. They took skin grafts from my right thigh to use on my right arm and right elbow."

In this recitation, body parts are jumbled together, as if the pain in his jaw were referred to his right elbow and arm. I do not ask to see the thigh from which the flesh was removed, though now that I know what to expect, based on what I saw of his arm, I am curious.

"They tried to rebuild my right eardrum. I lost about half my hearing," he says.

I try to listen as a doctor would, taking in the facts. But the look of pained innocence on this boy's face makes it hard to stay detached. I am irritated by a familiar feeling of wooziness, which I try to hold at bay.

"I need to warn you that I might ask you questions more than once," I tell him. "I might forget what you said five minutes ago. I didn't go through anything like what you went through, but I've been told that I have some symptoms of PTSD. When I hear a story like this, I begin to lose track."

I don't know why I feel the need to confess my difficulty maintaining control of my thought processes. I have not been asking many questions, so how would he know that I cannot follow? His story is coming out on its own, according to a logic I don't yet understand.

"That's okay," he says, grinning cheerfully. "I can't remember what I told you three seconds ago!"

I recognize this grin, the gallows humor of a man whose psyche is still partly in the grip of the threat of death. It isn't yet clear how much of his psyche death is prepared to return.

"How did your parents feel about your joining the military?" I ask. "How did they feel when you got wounded?"

In this moment, I am furious with the parents of this beautiful boy. How could they have allowed their son to join the military? Did they not understand the risk? Did they not understand that this war is a kind of class warfare?

"They didn't want me to join, but what could they do? I couldn't afford to pay back my college loans. I was over twenty-one. When I was wounded, they were devastated," he says.

He looks away, processing some private pain.

"When was this?" I ask, trying to impose a timeline on this narrative, if only to keep myself connected to historical time.

"This happened in April of '06. I had done my time. I was supposed to be free by then. Betsy and I were going to move to Canada. But then I got a notice: stop-loss."

"What does that mean?" I ask, stalling for time.

"Stop-loss means that you have an extension of your contract. At the request of the U.S. Army."

The word *request* throws me. "Does that mean you could say no?" I ask, although I already know the answer.

"You have no choice. You can go AWOL, but then they put you in jail.

"So Betsy and I find out, we're not moving to Canada," he says. "I'm going back to Iraq."

And when he comes back, I know now, he will discover that this wife he wanted to start a new life with was "bad." Instead of starting life in a new place, he will return to his hometown, a changed man, a man he barely recognizes as himself.

Once again I try to tether us to time.

"What year did you finish your degree?" I ask.

"I finished college. I signed a contract that said four years of active duty. I did that in June of '02."

I take note that he has to talk himself through time to discover the year, to get the sequence right.

"What happened after you signed the contract?" I ask.

"I went to basic training."

"For how long?" I ask.

"Two months."

"Then what?"

"Then I went to AIT."

"What is that?" I ask.

"Advanced individual training. That takes two months, too.

Then I went to airborne school in Georgia. That's where they teach you to jump out of planes."

"Did you like it?" I ask, knowing that skydiving was part of his dream, hoping for a moment of relief.

"No," he says. "But I didn't mind it—"

"But you wanted to parachute," I insist.

"You're crammed into a plane with a hundred other guys. You're connected to a line. You have your rucksack, that is fifty pounds. You have your gun. You're so loaded down, you waddle like a duck."

In my minds eye I see a duck loaded down with a gun and a rucksack. He dreamed of soaring. He wanted to soar like a bird of prey, not waddle.

"Then you jump. You're only eight hundred feet off the ground. They had told me, you'll be at ten thousand feet, jumping out of planes. It's nothing like they told me it would be," he says, bitterly again. But maybe not bitterly enough.

"How long did that take?" I ask.

I notice a slight frown on Erik's face. I cannot tell if he is annoyed at my insistence that he tell me this history in chronological order, or if providing the information is difficult for him.

"Airborne school takes three weeks," he says. "After that I was sent to Fort Bragg. You go through the reception area. They in-process you, do all your paperwork. You get assigned to a job. Whatever they need. I was a cook. I studied cooking. I got my associate's degree in culinary arts."

I notice that his sense of time is disjointed. Was there something memorable about this "in-processing" in the reception area of Fort Bragg that makes him recall that moment? Did he provide that detail to demonstrate his annoyance? I wonder, too: Did my insistence that he provide me a timeline make him think that I'm slow? Perhaps he doesn't remember that he already told me that he studied cooking. Or perhaps he doesn't know that it is the sort of thing I would remember. He doesn't know that I, too, love to cook. Running a café is one of my fantasy jobs.

"So I cooked down there," he says. "It was horrible. I worked seven days a week, sixteen hours a day. From four thirty in the morning until eight thirty at night. On top of that, you're staying in shape. It's really hard. Every six weeks you get a weekend off.

"And then I got deployed. They don't give you much warning. You have to inventory your equipment. You have to clean your weapons. Make sure everything is ready."

He goes on at some length about these difficult preparations. I grow impatient with his detailed recitation of bureaucratic requirements, with the military jargon.

I cut him off. "What did you do when you got to Iraq?"

"The first time I was in Iraq I cooked. You cook off of MKTs."

"What are those?" I ask. Once again I find myself impatient. Why all this jargon? Did this boy forget how to speak English? Perhaps, it occurs to me now, acronyms and jargon cauterize feeling.

"Mobile kitchen trailer. It unfolds into a Bunsen burner. Like you would use in the woods. But we have big military ones that run on jet fuel. Everything comes in packages. Mostly you just heat stuff up. But it's hard. It's hot. You've got to do all this with a gun on your shoulder. Eventually they built us a kitchen. Then we could cook burgers, hot dogs, pizza," he says.

"How long did you do that?" I ask.

"Seven months," he says. "You cook, but you also have to do the infantry stuff. Hunting for people. You could be in the middle of a burger. They will say, 'We're going to the base.' We always had our weapons on while we were cooking. That was annoying. You don't want to get it covered in food because it wouldn't work. But we had to be ready to be called up to the base, any time."

"What were the dates you were in Iraq?" I ask.

"First time August of '03 to April of '04. Second time August of '05 to April of '06."

"Do you think you had PTSD when you came back?"

"No, not the first time," he says. "Not the second time either."

"So it's April 2006. You're back from your second tour of duty, and you are on your way to Canada . . ."

"Yes. But then I was handed orders, stop-loss. The orders said they needed cooks. But when I got back to Iraq for my third tour, I didn't serve as a cook, I was in the infantry.

"Before, I was cooking for infantry guys, helping out the infantry guys once in a while. But now I was an infantry guy. Out doing door-to-door searches. I was doing something I hadn't signed up to do, wasn't really trained to do."

I am puzzled.

"I had only two months of basic training," he explains. "When you're a cook, you have to go to the range and shoot your weapon. You get called out to help. But most of the time you're cooking. I was an expert at the cooking part. I was mediocre when it came to the infantry part. The other guys were regular infantry guys. That's all they did, on a daily basis. But they needed people to fill some empty slots," he says. So that was that.

"The day I was injured, we were on a foot patrol," he says. "We were out walking. It's ten AM, my group's time to go out. I was the NCO in charge that day. That means I was the first guy to knock on the door. I was the first one to come around the corner. The first guy anyone would see. We checked a couple houses, no problem there. We talked with the people, tried to find where the bad people were."

I am very nervous now. Petrified of what might be around the corner.

"Where was this?" I ask, trying to anchor myself with latitudinal lines.

"I don't even remember where it was," he says. "Some hut in

the middle of nowhere. A bunch of little buildings with a river running through it. They needed our help."

Of course it was in the middle of nowhere, I say to myself. We are lost out there. It's a sea of unpronounceable names, where you can't tell the good guys from the bad guys.

"We were walking through the woods. Not exactly woods, but that's the closest word for it," he says.

I think of all the scary stories that take place in the spooky woods. "Hansel and Gretel." "Little Red Riding Hood." Stories my son once loved to hear.

"We came out of the woods to a main road. I told everyone to stop. There was a fifty-foot crater. Filled with water. They like to use those over and over."

What does this mean, over and over? I want to ask. I want to put the brakes on this story. I need to slow him down, to make sure there is time to see the scary thing around the corner. But I discipline myself to stay quiet. I can ask questions later.

"They set off bombs in the same place again and again," he explains, answering my unvoiced question. "I went up to see if it was okay. Everything was fine. I get up to turn around, to say to my guys, 'Let's go.' That's when someone with a detonator set it off. The bomb. It was buried in the ground.

"As soon as the bomb went off, the shock wave knocked all of them down. After that it was gunfire and screaming and yelling. Anything that isn't friendly, that isn't a U.S. soldier, is a liability."

His words are coming out chaotically now, associating from the sounds he heard, perhaps still hears.

"You get rid of that liability asap," he adds. He doesn't spell out the letters, as I would, but creates the word *asap* out of the acronym. "Instantly," he continues. "No one knew who had set the bomb off, so whatever you see, you just eliminate it."

Eliminate "it"? Is there some memory still buried, which he is not ready to reveal? I do not ask. Later, I will wonder if the memory of the aftermath of the explosion evokes in him the pain of some other event, some other "liability" that got "eliminated," or perhaps the opposite—a "liability" that did not get "eliminated asap," and an American soldier died.

"That's how you get collateral damage," he continues, with pained features. A wise boy. General Sherman's words come into my mind now: "War is cruelty, and you cannot refine it."

"Later I found out they buried propane tanks. They buried two of those under the road. . . . They can use anything to set them off. They could use a key like that one," he says, pointing to my electronic car key on the table between us.

My mind is so involved in managing this scene that I've forgotten I have a car key, or even a car. I am surprised that Erik is able to shift gears like this.

"Then the medic came over. Did his medic job. Stopped the bleeding the best he could. They called in a life helicopter."

I ask him to explain. "It's a Black Hawk with an ambulance. They medevaced me out to where the military surgeons were. They rushed me there, took me straight into surgery."

But he is not done with the scene of the explosion. I wonder if he will ever be able to expunge that scene from his mind's eye, to stop the tape of the sounds in his head.

His words are even more jumbled now. "When I was lying there right after the explosion. I can start to remember. The medic saying I'm giving you a dose of morphine. Not enough to put me out. Hooked me up with an IV bag. I guess to keep me from passing out from the pain. Still in a lot of pain. I didn't really understand what was going on. My buddy who was with me. I remember he was holding my hand. He said, 'We're gonna have a beer when I get home.'

"I was in Iraq in surgery, not even a day. They stopped the bleeding. Got me to the point where it would be okay to fly.

"A bunch of generals came. You know, the people who don't

do anything," he says, conspiratorially. "They tried to shake my hand, told me I did a good job. They don't even know my name. They say, 'Good job,' and move on to the next injured guy. 'Good job,' " he repeats bitterly.

"I can't see him. My jaw is wired shut, no one told me why. My eyes are swollen shut. I ask for a phone. I asked to talk to my parents. They called. I talked with my mom. I don't remember what she said or what I said. My commander flew in on a chopper to see me. Out to the base. Then I woke up in Germany," he says.

Once again his story is somewhat out of temporal order.

"They put me on a plane," he explains. "It's a cargo plane. They had me on a stretcher. There were nurses and doctors. I don't remember taking off or landing. The only thing I remember is the nurse saying you're in the operating room now. It's cold, the bright lights. They go, 'You're going to be okay.' "

"Did you think you were dying?" I ask.

"I wasn't happy. I knew something was wrong. But I didn't think I was dying. There were people like you. I knew I was in a safe place."

A sense of relief floods my nervous system. I have been so worried about retraumatizing this boy. I am moved, nearly to tears, that he sees me as a "safe place," even though I'm witnessing his recalling of the moment his life hurtled in an entirely new direction. He continues, apparently unaware of the effects his words have on me.

"Once I came out of surgery, they weaned me down. Off the drugs, the painkillers. That's when they told me what was happening. I had a dozen surgeries in four days in Germany. And then they sent me to Andrews Air Force Base and then back to Fort Bragg. Then I went to the University of North Carolina, in Chapel Hill, to get the skin graft. I was there for four days. By that point my parents were there, my sister was there. They stayed with me."

I breathe a sigh of relief.

"When was this?" I ask.

"The date of all this is March 27, '07," as if all these events occurred on a single day. "That is when I got blown up. I got discharged on April 13 of '07. After that I needed to get occupational therapy. So I went back to Fort Bragg. I spent a full year learning how to use my arm again."

I see now that the order of his recollections does not follow a timeline, but a system of referred pain. He is walking me from scar to scar.

Now a man walks up to our table. I have a vague recollection of a heavyset man in a red lumber jacket and gray sweatpants. But it is as if I were looking at him underwater. He walks right up to Erik. "I just want to thank you for your service," he says, respectfully. And then, more softly, as if he doesn't want to impose too much, he adds, "You're a great American." I watch the man walk back to the table next to ours to gather the leavings of his late-morning snack.

I had no idea that this man had been sitting next to us. For how long? I am following Erik's story so closely that it is as if I have become him. But I seem to be feeling more emotion in response to his story than he is. The words flow out of Erik like water flows, even if he gets stuck, occasionally, at bureaucratic dams. For me, there is friction, burning, pain.

The appearance of this lumber-jacketed apparition reminds me that I am not in Iraq, not in a hospital, not training my mind to communicate with the muscles of my arm. I feel a great sense of relief to remember where we are, to see that at least one American is prepared to thank this young man.

A small smile plays on Erik's lips. He seems to allow himself to feel the faintest possible fleeting pleasure. But in less than a second he gets right back to work, to the story he is ready to impart.

"I can feel myself touching myself, here," he says, rubbing the meat of his right arm, which is now safely covered by a moss-green sleeve. "But not like this hand," he says, touching the other arm. "You lose sensitivity.

"I got on the list to see a shrink. When you go through the MEB [medical evaluation board] process, it takes a long time. I wanted to get evaluated for PTSD. I had to wait for a full year.

"By that time, it's really too late," he says. "Things get deeper and deeper. It's too late," he says, bitterly.

No, it's not! I want to shout. Somehow, I see the ember of anger in the corners of his eyes as a very hopeful sign. He is not resigned, I am relieved to see. He is angry, even if not angry enough. Suddenly it seems to me that anger can be a lifeline.

"A friend of mine who had gone over in a different unit, he lost a quarter of his skull. He would say to me, I notice that you've changed. He noticed things about me that I couldn't see about myself. I used to be, like, la-di-da about everything. But after my injury, I was easily aggravated. I got angry all the time. I couldn't stand to be in loud places or congested areas. Even something like this," he says, pointing to the nearly empty coffee shop, now that the lumber-jacketed apparition is gone. "I wasn't like that before. Before my injury I liked to go out."

I ask him whether fluorescent lights bother him, wondering if my strange aversion to these lights is more broadly shared.

"Bright light bothered me a lot. I had extreme sensitivity to light. That went away," he says. I find that hard to believe. He doesn't like being where he is right now, in the light.

"But there are still some things I can't handle. Multitasking," he says. "Forget it.

"They knew I would be medically discharged because the surgeon who worked for the military filled out the paperwork. 'This soldier has been hindered due to wartime accidents,'" he says, apparently reciting official language. "That means they should be responsible for my medical care. It means that if my arm starts to fall off, I can go back to the government and get it sewn back on. But I had a brain injury. Studies show when a bomb blows up next to your face, you have issues. That is why I started the MEB process. There is only one shrink at Fort Bragg who is authorized to evaluate us for PTSD. There were six hundred people all waiting to talk to him.

"It's retarded," he says angrily. "You wait and wait. You have no family and no friends there. My unit was still in Iraq. They let me go home for two months to be with my family, to heal. But then I had to go back to Fort Bragg, to sit and wait. They still make you come to formations. I couldn't sleep. I was having nightmares and flashbacks. They gave me drugs for pain, drugs to sleep.

"I was wondering if I would live," he says, a statement I find startling. When he really might have died, he wasn't afraid. But depression and this state of nothingness he felt, which he calls boredom, were nearly unbearable.

"I was stressed, and I was alone. I don't think this is a good thing to do to soldiers with brain injuries, make them wait that long to see a shrink. But I didn't have any option. Once you start the MEB process, there is nothing you can do but wait. It's mentally torturing people. This went on until April of '08. Finally I get to see this shrink."

"How long did he talk to you?" I ask.

"Half an hour," he says. "He tells me that I'm fine, that I'm stressed out because of the MEB process. His recommendation was 'Deal with it. You're not the worst of the worst. You're fine.' I'm having flashbacks, and I'm having nightmares, but he says, 'It sounds like you're anxious to get out of the military.' "

I ask what he means by flashbacks.

"If I'm in a crowded area and a bang goes off, I'm jumpy. I was having nervous attacks. I wasn't diving behind things, so it could have been worse," he explains. "I'd be out with a couple buddies.

If there was too much noise, too much going on, I would just say, I'm leaving. I'm done. I need to be quiet and alone. That is what I learned to do. I knew to leave a crowded area. I knew to go home."

This need to avoid crowds—I know all about that. But I was a shy child, so it was easy to attribute my own dislike of crowds and noise to a personality quirk—and it probably partly is. Still, I am not ready to accept that we have the same syndrome, the same changes in our brains. His symptoms are so much more understandable, so much more justified. His story is so sad, while mine, it seems to me, is all about shame.

"Road rage," he says. "People were always in my way. I yelled at people, cursed at them, 'Get the fuck out of my way.' So easily irritated . . . I was never like that before. I would honk at people, shout at them, 'Why the hell are you going so slow!'

"I can be at work. Someone will say something that I think is ridiculous. I'll say, 'Man, you're an idiot. . . .' I just can't deal with things the way I used to. I get stressed out, anxious. Nervous."

"What do you dream about?" I ask.

"In all of my dreams, everyone dies and I'm the only one who lives. Or vice versa. It's always the same. Two weeks ago I dreamed I was on a boat. Everybody falls overboard and they get chopped up by the propeller and I live. . . . Another one, I'm at the bottom of a hill and a boulder is rolling toward me. I'll wake up drenched in sweat, my heart rate going through the roof. I'll be so thirsty. I never had dreams like this. Very violent. He was wrong," he says, referring to the psychiatrist.

"So you told the psychiatrist all this, and he said you didn't have PTSD?"

"Yep. I talked to him for half an hour, and that is what he said."

"Did the psychiatrist prescribe medication?" I ask.

"No," he says.

I am uncertain what to say. Should I reveal the horror I feel at hearing this story? Would that make him feel better or worse? Is this boy's body and mind considered expendable? Is a soldier's spirit, which he put on the line in the service of the country, too expensive to repair?

"The worst is, I get really spacey. I can't focus. I can't talk on the phone very long. I start to get light-headed and dizzy. If a conversation lasts too long, I'll start sweating and get stressed out. I forget everything. I have to have yellow stickies everywhere. That is why I needed you to e-mail several times about when and where we were going to meet."

I, too, cannot bear speaking on the phone. I, too, get light-headed and dizzy if a conversation lasts too long. I am astonished to hear all this.

"Do you get lost a lot?" I ask, wondering if this personality quirk of mine might also be a marker of trauma.

"It's not that I get lost more easily," he says. "But if I get lost, I cannot find my way back. I just don't know how to handle it. I don't even try to multitask. I can't do that anymore. If the phone rings and I'm doing something else, I won't answer it. At work, I can only handle one order at a time. When it gets really busy, they know I can't handle it. That affects my job. They know me and understand. They can look at me, they'll know, Erik is down."

I tell myself that the right thing to do is to tell Erik what I think.

"You need to take all this seriously," I tell him. "PTSD creeps up on you. It can take years before you realize it. You need to get them to help you."

But I can see, from the glaze in Erik's eye, that this is all he can take. "Are you dizzy now?" I ask him, gently. "Should I take you back?"

"Yes," he says. "I feel a little light-headed. I'm starting to sweat.

I don't know why. We thought it was because of the medication, but it turns out it's not. I was looking at you earlier—and I noticed that the shape of your face got fuzzy."

He will rest, he tells me, before driving back up north. He got one of those GPS systems, he tells me, to help with his problem of getting lost.

I know that the biggest danger for me, after an interview like this, is making it back home without getting lost on the road. I haven't yet bought one of those GPS things.

I take stock of my mental state. I am a bit fuzzy-headed, but I tell myself that once I get to the highway, it's a straight shot home. I rejoice, as if I've run a marathon, when I make it to the highway. No mishaps! But after several miles, I hear a tone indicating that the car will soon be out of gas. I get off at the next exit where there is a gas station nearby. I know, of course, to discipline my mind to memorize the route from the highway to the service station. Right at the exit, then up a small hill to the left.

I am in a small town with one of those gas stations where you fill your own tank and pay afterward. I insert the nozzle into my car and wander up to a low-slung building where I will pay once my tank is full. There is one cashier and a long line.

The buzz of the fluorescent lights penetrates my skull. I push the sound away. I notice that the cashier wears a name tag that says "Ahmed Khan," just like the name of a friend of mine. But unlike my friend, this Ahmed is Indian, not Pakistani. As I wait, I think about what it means to be a new immigrant. I feel a wave of love for my country. Then I wonder how Mr. Khan was treated after September 11.

I observe the customers waiting in line. My attention is drawn to a young father whose daughter is already drinking her Fanta while her father waits to pay for his gas and a Coke. Caffeine, I think to myself. Good for you when you drive. I decide to make myself a cup of tea. I pour water from the urn, put a Constant Comment tea bag in the heavy paper cup. When I notice the hot tea burning my fingers, I put a sleeve on the cup.

The person in front of me in line is an overweight lady wearing sweatpants. I am too old to wear sweatpants, I tell myself. I resolve not to wear sweatpants in public anymore, but of course I know this resolution won't last.

"Coffee is a dollar forty-nine," Mr. Khan tells the lady in front of me. She is already sipping her coffee; I can smell the pleasant scent of hazelnut-flavored coffee. After I graduated from college with a chemistry degree, I considered taking a job with a company that manufactured flavorings. You could smell the factory from far away. I had seen this woman earlier at the urn, considering her options, adding two packets of Equal to her coffee.

Fake sugar collects in your eyeballs, I think to myself. I have a vague recollection of a study at MIT, but I am not sure—does sugar collect there, too? My imagination doesn't stop at her eyeballs. I think about perfectly formed crystals of Equal lodging themselves in her arms and thighs. I see, in my mind's eye, tiny castles, glimmering like jewels, miniature replicas of the Bundt cake I plan to bake tonight for my son's birthday party tomorrow. She sighs and hands Mr. Khan a five-dollar bill, making it clear with that sigh that Ahmed Khan is charging her way too much. Foreigners. I imagine the castles, emerging through her clothing, becoming more spectacularly turreted with each sip. Mr. Khan hands her the change, which she counts theatrically, to make obvious her concern that Mr. Khan might rob her. What would it be like to walk around becastled like this? I wonder. Now it is my turn. I sign the credit card receipt and return to my car, prepared to make my way back to Boston.

I get in my car and drive off. I am puzzled by a loud crash, the sound of an M16 clattering to the ground. Well, of course it wasn't an M16. Still, there was that sound. Calmly, I stop my car and get out to investigate. I see that I forgot to remove the hose from my gas tank, that the handle has come loose from the hose. I walk back to the low-slung building to confess my crime to Mr. Khan. There is still a long line, but I interrupt his transaction.

Mr. Khan instantly shifts gears. He phones a person whom I imagine to be his boss, while irate customers wait. I hear Mr. Khan speaking rushed Hindi to an older man. I can hear the older man's loud voice.

Mr. Khan becomes visibly alarmed by whatever it is that the older man has told him. His English becomes nearly unintelligible. "Your good name," he stammers. "Here," he says, handing me a scrap of paper and a pen. I write my full name. "Your domicile." I write my address. "License. And birth date," he demands, as if wishing to humiliate me, to punish me for my part in the terrifying response of his boss. I do as I'm told. "Mobile here," he says, pointing to the only spot left on the small scrap of paper. I give him everything he requests, plus my credit card. I urge him to write down the number, telling him I will pay for any damage. I slink back to my car, suddenly aware that the interview was harder for me than I anticipated, harder than I realized even after it was over. Somehow, although it is a very simple route back to the highway, I get lost, distracted by the heavy feeling in my limbs, the familiar feeling of shame. Shock and shame. Such a familiar combination of feelings for me. What shocked me the most was how familiar Erik's complaints were. He is suffering from being like me.

I will talk to more soldiers after Erik. The stories break your heart. But they are for another book. I learned what I need to know from talking to Erik.

chapter thirteen

Faith

I have given my father the transcript of our interview so that he can make corrections. He has many objections, he tells me by phone. There is a tension in his voice that makes me consider the possibility that I should stop writing, or at least never publish this book. I am so deeply saddened by the burden of having to be the bad daughter.

I wait until he has read the entire manuscript. Then I wait until he is ready to talk. Weeks go by. I am on tenterhooks. During this dry period, we are in contact mainly through my son, who visits his grandparents every week. When my father is ready to talk to me, we agree to meet at his house so that we can go over his concerns in detail.

Both my father and I have prepared ourselves for this conversation with notes. He has a list of his objections to my draft; and I have a list of my proposed improvements. I notice him clutching a piece of paper. I think that he is sad. He is controlled when he feels most vulnerable. It unhinges me a bit to see him like this, but I urge him to begin.

"I am personally affronted by your characterization of my behavior after your rape," he reads, holding his sheet of paper at an awkward angle. Perhaps he needs new glasses.

His eyes stay down as he continues to read. "I thought our conversation about your rape was private. I did not know you were going to put it in a book," he says. "I am a private person. I don't want to publicize myself."

The way he enunciates "your rape" cuts at my heart. In the moment he says it, I have the feeling that he wants to distance himself from an ugly, lewd thing that happened to me, resulting from a flaw in my character.

And yet, I am certain that my father does not mean to be unkind. On the contrary, I finally understand that my father comes across as cold and dismissive when something hurts. It is not because he feels too little, but because he feels too much. Nonetheless, I feel myself begin to float, as if my spirit were about to fly out the window while the rest of me sinks into some dark shameful place underneath the surface of the earth. I will myself to stay in the room with what is going on between us, to hear and feel whatever it is that my father wants to say to me.

I notice the sunlight falling on the stone wall in the back garden. A tufted titmouse lands with a soft *thwack* at my father's carefully tended feeder, which he has engineered to maximize the access of birds while minimizing the access of squirrels.

"You agreed to be interviewed," I remind him. When these words come out of my mouth I recognize my defensiveness, but I cannot stop myself. "I told you I was interviewing you for my book, but you didn't believe me."

He doesn't respond to my interruption, but continues with his script.

"You painted a one-sided portrait of me. You made me out to be some kind of Nazi."

That word! I have hurt him. I knew this when I came here, but the shock of that word slaps me awake. I will have to bear my father's pain. How is it that he hears me this way, when I wanted to show him as a person harmed by the Nazis?

He looks up now to object, "There is no indication of how difficult you were, how many times you ran away." He is speaking with the hypercorrect intonation you can hear in his voice when he is agitated. I am afraid of my father's rage, but even more afraid of his pain.

I did keep running away. I was "difficult," as my father says. But why does he continue to blame me for this? Wasn't he partly responsible? My father's new home, which came with a new stepmother, did not feel like mine. I felt unwanted, and perhaps I was. So I ran away. There was that night in the graveyard. Other nights, too, always outside. I remember, now, a night I spent in a birch forest at the top of a hill, a forest that I loved during the day, at Punkatasset, a town forest. I used to ride my bike there after school and then find a sunny spot to read. But it was dark and cold that time I arrived at night, with shadows cast by the moon. I was shocked to find myself actually petrified. Petrified, I suppose, of my own shadow. But I awoke at dawn to the familiar wood, now brightly lit, that euphoric white light. And then I forgot the whole thing, until now.

He looks down at his list again. "There is no mention of your disruptive intransigence."

Later, I will realize how much this dismissive phrase hurt me. Must I take the blame for all the upheaval of those difficult years? Did no one else in our family bear any responsibility?

I tell my father that my "disruptive intransigence" did not come up in our interview, and for that reason those words are not in the text I gave him to read. I explain that I have done my best to stay true to what he said and what I said that day; that although this is not a work of social science, it is a habit I acquired from more scholarly work. I promise to make sure that readers learn about my disruptive intransigence.

I remember whipping myself up into a mutinous rage when I was thirteen years old, repeatedly telling my father and the whole world that I hated him. This act of insurrection was inspired by my father's seemingly arbitrary decision to remove my sister and me from Lisa's home, where we had stayed for two years, even after she and my father divorced. I blamed him—wrongly, it turns out—for wrenching me away from my second mother, less than ten years after I had lost the first. I could not accept and did not want to live with a third mother, so different from the second one, and so different, too, from what I remembered of the first. I sensed and was terrified that the feelings I had about taking on yet another mother were reciprocated. I was argumentative and angry, "disruptively intransigent," though inside I felt abandoned. Something seemed unnatural about the situation I faced—too many mothers, too much shifting of beds and homes at what felt like my father's whim. When we arrived at our new home, I would not sleep in a bed, but instead, slept on the staircase. I do not remember how long my insurrection persisted; my memories of past pains are hazy now, and unlike the rape itself, there are no police records to consult.

My father did not respond to this unprecedentedly mutinous act. He remained impassive in the face of my shouting, and let me sleep on the staircase. At first I was relieved. But then I grew alarmed. My father had been in the navy. Was no one in command of this ship? No one prepared to squelch my mutiny? I realize, now, that my father must have been utterly confused as to how to proceed. It was Lisa who wanted me out of her house, but my father couldn't tell me that. She had set some conditions for my continuing to live with her, and my father found those conditions unacceptable. He wanted to protect me from the truth. Although he didn't tell me, I understood one thing, at least: I was unwanted, and rightly so. I was a difficult child, for reasons that no one understood at the time, least of all me. My young second mother had given up my father, who was significantly older than she, for a younger man, some ten years older than I. She and her young husband wanted to build their own family, without the detritus of my father's past life. So there we were in this difficult time—each of us wounded, half blind with pain, wounded and wounding. My wounding took the form of a shouted insurrection. The grown-ups, at least in my father's home, used covert methods—smiles and denials. But I was the sort of child who felt her way through the world, and the damage was done. It was this capacity to feel too much that made me intransigent. It was not a good way to commence one's relationship with a new stepmother.

Once again, my eye is drawn to my father's bird feeder. Chickadees are so common that I rarely pay attention. But if you look closely, such beauty! The black cap and scarf, the white face, the elegant tail. My father spends a lot of time caring for these birds. He is building a birdhouse now with my son. They have drawn up architectural plans, and are planning to begin construction next week. What an extraordinary grandfather he has turned out to be.

He continues.

"There is no acknowledgment of how much Gwen [his wife] and I tried to help you during your troubled childhood. How much we have tried to help you in your troubled life."

The words "your troubled life" are so cruel, and so bizarre, that I find a way not to react. I will feel about this later. No wonder I wanted to sleep in the woods, or among the dead. Perhaps he actually means "troublemaker," or "troubling."

I don't know if he fully realizes this, or the power of my love for him.

I do not believe that my father truly believes that my life could be summarized as "troubled" any more than anyone else's in our family, including his own. This period of insurrection, after all, occurred nearly forty years ago, just before and immediately after the rape. But my father's recollection of this period of our lives, and my intransigence, seems to haunt him every bit as much as it haunts me. And now I wonder, Does my father believe that I am judging him for not coming home to us right away? Morality is beside the point, I want to say to him; you made a rational decision, uninformed by your emotions. I've done that, too—so many times. There is a reason that God gave us feelings, as annoying as they often are. Emotions guide us, every bit as much as thought, even though they would seem to muddy rationality. I had consulted a therapist to escape from feelings—but now, it seems, I've discovered their utility.

I hear the drum of a woodpecker. This sort of tapping-ticking sound is hard for me to bear in moments like this. I have learned to recognize that when a tapping sound enters my consciousness harshly, like this, I must be distraught.

"A downy woodpecker," my father says, pointing to the wood on the far side of the house. "He lives over there."

He continues with his critique, not just of my transcript of our interview but of my character.

"You lack social skills. You have no social graces."

There is no point contesting this description, as I know it to be utterly on the mark. For a moment I see myself completely alone, abandoned by friends and family, pushing a shopping cart with my few remaining belongings. I hear the words "Honesty is the best policy." Is it really the best policy? Or is it just that I am too awkward, too lacking in social graces, to be a nice person? My honesty, no doubt, is what has led to my "troubled life."

I recover my equanimity. "Has it not occurred to you that we are similar?" I ask my father, laughing. "That I might have acquired this trait from you?"

I see his lips move, a bemused grin, but he is not quite ready to concede the point in full. He wants to differentiate his style from mine, or at least justify himself. He tells me that his "propensity to speak the truth" is common for scientists of his generation. "It's the culture I grew up in," he says. "It seems ridiculous to dissemble."

For my part, my propensity to speak to truth led me to say something so rude at dinner last night that shame roused me from sleep in the middle of the night. I was asked if I liked the wine. I did not. Why didn't I just say that I did, like a normal person?

Perhaps it has to do with having grown up during the Depression, my father hypothesizes. Or coming of age during World War II. He tells me that he recognized himself in a paragraph in Richard Feynman's autobiography, in which the physicist says something about the irrationality of lying because the truth always reveals itself in any case.

He is especially troubled, he tells me, that I quoted from an e-mail he sent me fifteen years ago, which I saved and which I intended to quote from, in which he explained what he was doing in Norway and why he did not want to curtail his trip.

"I was responding to your pointed questions," he says, by which I think he might have meant leading questions. "And I did not give you permission to quote me. I see this as a betrayal of trust," he reads, as if laying the groundwork for a lawsuit.

I feel a hollowness in my stomach at the thought that my father might be planning to cut off all relations with me. Will he, once again, "move on," live for the future, leave the painful past behind, abandon me? But there is something incongruous about my father's delivery. He seems to be playing a role, or perhaps that is my wishful interpretation of his puzzlingly gentle demeanor.

He turns to our mountain-climbing trips. Now there is passion in his voice. I understand now that it is the discussion of our climbing trips that troubles my father the most, more than anything else he discussed. He looks up from his paper.

"I did not go climbing because I wanted to endanger myself or others. I wanted to meet a challenge! It seemed incredible to me that a person could be comfortable when it is below zero. I wanted to experience that," he says.

I've been in subzero temperatures and I find it quite cold. Somehow, despite my father's best efforts, I grew up to be "candy-assed," at least in regard to cold. But of course I know what he means. If you walk fast enough on a steep trail, you will remain warm, even at below-freezing temperatures.

He looks down at his list.

"I learned how to be a leader—I'm a qualified leader of groups in the mountains." I tell him that of course I know he is a certified guide.

He continues reading. "I like to meet challenges—including talking to you about this now."

I smile. "I agree. You are very brave," I say.

My father's list of complaints about my character suggests that he might be prepared to disown me emotionally, as he has before. I was, no doubt about it, a difficult child. My father admits that he gave up on me entirely when I was a teenager. He hasn't told me, but I know it to be the case, that he gave up on me more than once. Still, I don't believe that he means to abandon me this time, even if I fear that he may have been toying with the idea.

"How many trips did we take in the mountains? Can you imagine what it was like for me to take a car full of teenagers up to the mountains again and again? Why didn't you talk about the other trips? There were tens of them, perhaps a hundred. Why did you select the least pleasant ones to talk about?"

In fact, I never really thought about this. Three out of four of the teenagers in our household—my stepbrother James, my sister Sara, and I—always went on these trips. But my father would bring any of the neighborhood boys who wanted to come. In my father's view, mountain climbing builds character, and he wanted to offer this gift to as many kids as he could, as many as he could fit in the car.

After my father retired, he joined a National Science Foundation–sponsored program to bring scientists into the public schools. He started volunteering in a public school in Lowell, teaching physics to the mostly Cambodian kids who live there now. Physics is what pulled him out of the German-Jewish ghetto. Physics is what rescued him from a life of being a victim. Perhaps it would work for these Cambodian kids, too. He spends a lot of time devising new ways to show the children how gravity works, conservation of momentum, the impact of friction. He jumps up on tables, sets balls in motion. He tries to repeat Newton's and Archimedes' and Galileo's experiments. "They are simple and accessible. And they don't require fancy equipment, which we can't afford," he told me.

My father once explained to me that for an observant Jew, anonymous philanthropy is considered a greater mitzvah. The recipient might be humiliated to encounter his donor, so the relationship should be anonymous, and at arm's length. And if your good works are motivated in part by your desire to be recognized by the community, they count less in God's eyes. No one in Lowell knows my father; his name doesn't end up on a program or in a yearly Christmas letter. His work in Lowell comes very close to pure altruism. Pure altruism, for me, would have to involve no pleasure for the donor, not even the pleasure of pleasing God, and certainly not with the hope of receiving rewards after death. I am pleased to say that my father gets a thrill from demonstrating Newton's laws with books and balls.

Volunteering, I always thought, is what ladies who lunch do

to fill up their time, something I think of as "girlie." It is better, and more rational, if you care for the poor, to spend your time working for money, as more often than not, a check does more. But recently, I have found myself volunteering in my son's first-grade classroom. It occurs to me that my father's work in the Lowell public schools has made volunteering seem more manly, and therefore more acceptable.

"I stuck to the transcript of our conversation," I remind my father, "and those other trips didn't come up." But then I ask his permission to take my turn, to explain how I plan to put my interview of him into a broader context.

"Okay," he says. He puts his slip of paper down on the table. He sees me eyeing it. "I'm not going to give it to you," he says, teasing me for my curiosity, but also slapping the hand that would touch the fire.

I pick up my list. As it happens, my list begins with our mountain trips. The first item is:

1. Positive aspects of climbing with my amazing dad.

I read this sentence out loud, but skip the childish word *amazing*. I'm already feeling like an admonished child, a child who is having an absurdly hard time growing up. But also I feel shy. I can say nice things about my dad behind his back, but it is not our style—neither his nor mine—to say anything complementary to the other's face. It might sound like flattery. It might be construed as manipulative.

Here is what I did not say, and could not say, to my father in that moment. Those mountain-climbing trips saved me from despair. They changed my life, made it possible for the best in me to emerge. Yes, climbing taught me discipline. But more important, it opened me up to a capacity for bliss.

Even now I can rest in the memory of allowing my whole

body to sink into a bed of moss by a mountain stream. It's been years since I laid myself down into a bed of moss, but the very thought of that soft bed and the moist, peaty scent brings back a feeling of rapture. If you looked closely, you would see that the ground covers—the hair-cap and peat mosses; the club mosses and horsetails—comprised a forest in miniature, mimicking the taller trees above. It was as if God wanted to create a separate forest especially for children.

In those pre-giardia days, you could drink from the streams. There was a slightly metallic taste to the streams in the spring, the scent of melting ice and snow. I remember, too, the sour thrill of wood sorrel on my tongue. I liked to chew it as we climbed, like gum.

In the spring, all was mud at the bottom of the mountain. As we started our ascent, a thick deciduous sludge clogged the trail as well as one's mind. The blackflies formed thick clouds of itch. But all that black and gray and itch made the burst of color when you came upon spring's first wildflowers even more astonishing. I can still bring into my body the joy I felt at seeing the first trillium of spring, which seemed to be telling me, "Never give up hope, spring will come." My breath would catch at the sight of violets—so common in the woods at home, so surprising in the mountains. The violet's message to me was "Keep up your courage, stay true to what you believe in." There were bunchberry flowers in tight white clusters, feathery white sarsaparilla flowers, false and hairy Solomon's seal. I took a special delight in knowing the names of these flowers, having taken a nature appreciation course, odd and troubled child that I was. There were the sturdy, reliable flowers—marsh marigolds and starflowers and wood anemone. And the more delicate ones. If you were lucky, you might stumble into a perfect circle of lady's slippers, obviously left by fairies. The incongruously intricate bells of the wild columbine were the biggest surprise of all—the outer petals red, with upward projecting spurs; the inner petals a pale yellow; the delicate filaments and anthers a brighter, bolder, joyful yellow. Why would God place such an elegant flower in the deep woods? Other flowers came at the end of the summer, but by then the winter sadness had already dissipated, and the effect of the blooms was not the same.

As we climbed higher, the flowers became fewer. But there was the comforting, slightly tannic scent of warm, dry pine needles, which bit the nostril the way tea tastes on your tongue, making you feel excited and clean. The aroma of balsam shoots when you came to a shady ridge brought on a feeling of euphoria, even among those depressed by the prison of their own disruptive intransigence. In the background of all these scents was the odor of my father's mountain sweat, the mingled scents of joy, hard work, and Woodsman insect repellent. You could rest your soul in that smell. To me, it was the scent of safety.

Best of all was the thrill of "getting your mountain legs." That was when your muscles ceased to ache, when your body felt light and free, and you could run up and down the trail without seeing what lay ahead. This was bliss, knowing that your body—so unreliable in other circumstances—would react automatically and appropriately to a boulder or a fallen tree or a sudden turn in the path.

By the time we rose above the tree line, we would all be feeling drunk on the mountain air. The unencumbered breeze blew away my winter angst, making me feel hollow, even transparent, opening a space for ecstasy. During our descent, when our legs were beginning to tire, my father would cheer us along by playing his harmonica, or sing to us with his great baritone, the vibration tingling my spine.

The feeling of euphoria stayed with us a day or two when we returned back home. But the memory of this rapture is what kept us returning. It still keeps me going, even now.

We changed up there, my siblings and I, and so did my father. A great burden was lifted from my shoulders—the burden of my own badness, and the burden of my father's pain. His voice took on a new, lighter timbre. There was a contagious euphoria to his gait. I could experience him looking at me with his soft brown eyes, rather than just toward me. His sadness didn't leave entirely. I could hear it when he keened to the mournful call of the wood thrush at the end of day. But in the mountains, I was forgiven. And my dad was forgiven, too.

I continue reading through my list for my father. I stumble over my words, painfully aware of the awkward asymmetry of my mouth, the way my lips don't meet in a neat straight line. Somehow I have reverted to an awkward teenager, trying, once again, to grow up.

2. *I love my father very much—I need to make that more clear.*

With this sort of mushy girlie talk, it's best to look away, so I do.

3. *My father experienced trauma.*

I look up briefly, to see whether he will object to this word *trauma*, which is awfully girlie, and certainly smacks of "examining one's navel." But he does not object, so I continue.

4. *My father had or has post-traumatic stress disorder.*

I pause, certain my father will admonish me for using this silly word, this "psychobabble." I don't look at my father, but down at my list. I wait, as if sniffing the wind. Remarkably, I sense curiosity rather than condemnation. The sensation of my father's curiosity—real or imagined—propels me on.

5. *My life would have been very different if there had been a book I could have read that would have explained that other people experience the changes in mood that I experienced—the hypervigilance and hypovigilance.*

"Remember that kids called me a space cadet?" I ask, looking at my father. He is puzzled. "I wish I had known then that I was dissociated, that it was a symptom of trauma. I am hopeful," I tell him, "that this book will help others."

"So your goal is to help other people with this disease, this post-traumatic stress disorder?" he asks. This is something he can bear. This is something that he might even be able to respect. I tell him yes, that is one of my goals.

But I want to keep going with what I came here to say. I look down at my paper and see that the next item on my list refers to my mother's death. I need to keep looking down, in order to speak the unspeakable.

6. *My father could not face my mother's death because it came after too many earlier traumas. New traumas reignite older ones, so that my mother's death, for my father, brought back the memory of the Nazis and of having to leave his home. My pain would reignite my father's pain. Therefore, I could not face my mother's death.*

I glance up at my father from time to time. He is not denying any of this trauma stuff. I have brought my mother's death into the room, and nothing has happened to either of us. We are still speaking, still looking at one another.

"Your grandmother would not have liked the way you spoke

about your grandfather," he says, surprising me now with a new mode of attack. My father has often made it clear that he thought my grandfather was a lecherous liar and a cheat.

"I thought you loved your grandfather!" he says.

"I did love my grandfather," I tell him, softly. How can I explain that one can love one's abuser? How can I explain that I know that my grandfather, my potential-maybe-almost abuser, loved me?

The topic of my grandfather, and what I have not quite said about him here, reminds my father of something he has never told me before, and which I never knew.

"He lost his hospital privileges," my father says. "Around the time that I married your mother. I never knew why."

My father wants to defend my grandfather from potentially false accusers. Even those we despise should be innocent until proven guilty. But that doesn't mean he won't share information that might undermine this defense. My father is an advocate only for the truth, even—or maybe even especially—when the truth hurts.

What would a physician have to do to lose hospital privileges in 1955? Those were the bad old days, when doctors were gods. Screwing your nurses—apparently a common practice for my grandfather, at least according to my grandmother—would presumably not interest hospital officials back then. What could my grandfather have done?

I ask Jack, my RA, to try to find out. But the hospital's records were lost in a fire, it seems. And there were no ethics committees back then.

I cannot say what my grandfather did to me, other than leer revoltingly at "beautiful ladies," saying "Oo lah lah"; other than comment, inappropriately and disgustingly, about my growing breasts. I cannot remember. He's dead, and cannot prey anymore on other little girls. I cannot even remember now what I know I once recalled about what happened in the shower when I was eight or nine. I cannot bear to remember. There are no photographs or crime reports or police files to consult, as there were in regard to my rape. And even if I could remember, I would not trust myself. I know that memory is suspect.

But a strange clue did come to us, in the mail.

Several years ago, shortly after my sister and I persuaded our father to talk to us about our mother for the first time (warning us, prematurely as it turns out, that it would be the last such discussion), my mother's best friend sent a letter to my father. She had found in her files a letter that my twenty-seven-year-old mother had sent her in January 1961. This friend lives in Israel. She sent a copy of the letter to my father, and my father then made a copy for me.

As my father handed me the letter, I had many conflicting feelings. A kind of nausea. I had always felt that my mother was somehow out of fashion in my father's home. She had a kind of old-fashioned look, not surprisingly, in the old photographs that my sister and I had seen at our grandparents' house. She seemed uncool. Curiosity about her was socially unacceptable in the atmosphere in which I grew up. I had somehow absorbed the notion that it was unhealthy, a kind of perversion. I was too ashamed to be curious or excited about seeing the letter, and the shame turned into light-headedness.

The letter is typed. The style is jaunty and optimistic, although my mother had been diagnosed with cancer a year earlier, and had recently undergone a bout of chemotherapy.

The letter begins innocuously enough. She refers to Kennedy's recent victory in the 1960 presidential election, and his appointment of his brother as attorney general. She mentions the shock of David Ben-Gurion's revelation, in December 1960, that the Dimona complex was in fact a nuclear reactor, which Israel had earlier denied. She mentions her recent appointment as chairman of the Foreign Policy Committee of the Syracuse League of Women Voters. She described her work with the league as a respite from changing diapers. The four of us—my father, my mother, my sister, and I—were living near Syracuse at the time, which my mother refers to as "the wilderness," apparently preferring to return to Manhattan, where she and my father lived when I was born. She survived the bout of chemotherapy, which "appears to have done its work," she tells her Israeli friend. "With luck, we've seen the end of that for a while," she writes.

Then my mother talks about me. "She is very feminine, and insists on wearing dresses and stockings all the time," she writes. This is a surprise, since I became a tomboy. "My parents gave her a rocking horse for Chanukah which she promptly named Misha the pawnbroker. God knows where she got that one from," she writes, amused. But then she describes a puzzling scene. "My parents were here last week and one evening my father was dressing her for bed—she opens her legs wide and pointing to the crucial area said to him, 'Touch me there, Grandpa, it feels good.' " She concludes that Jessie "talks a blue streak and is a real little sex pot." I was not yet three years old. My mother would be dead within a year.

Of course my mother didn't know, could not know, that child-care workers would eventually be trained, many years after her death, to look for precisely this sort of sexualized behavior in young children as a potential indicator of sexual abuse.

My father and I break to take a walk, into the woods that we both love. We speak about our neighbors—who is still here, who moved away, who died.

On our return to the house, my father says, "You want to see yourself as a Holocaust survivor once removed." He says this disapprovingly, making it clear that he believes I have taken on a dramatic persona unrelated to my actual experience, and that he finds this sort of pretension unhealthy and unethical. Most of the time my father does not see himself as a Holocaust survivor. And I, in compliance with my father's unstated wishes, have not seen myself as the daughter of a survivor, at least not until now.

"It's not that I want to, but that I am," I say, pushing against the tide of my father's disapproval. I can hardly believe that I managed to utter these words. "But in my case, that was just the tip of the iceberg. There were so many traumas. Our mother's death, your terrible marriage to Lisa, the rape." I say these things in part because they are true, but also because I want to relieve us both of the image of my father as victim, which neither he nor I can bear.

Still, difficult child that I am, I will no longer comply with denial. The cost is too high. I am not going to bear this burden alone. "Are you actually prepared to claim that your experience of Nazi Germany had no impact on you? That it didn't change you at all?" I ask him, still defiant in my "disruptive intransigence." How I wish he would know me.

There is a pause. "It made me more appreciative of life," he says. "I am like a person who has had a near-death experience. Everything is extra bright in my eyes."

This I know to be true. It's true for me, too. There are positive sides to facing down one's death relatively early in life.

When Chet first came back into my life, I was rude. Even more rude than usual, even for me. "How old are you?" I demanded to know, confused that he did not look much older than I, even though when I was a teenager, he was a grown-up. I thought he was around my father's age, which would have put him, at that time, at age seventy-nine. He wouldn't divulge his age, but he announced proudly to everyone within earshot that he had known me as a child. A mutual friend had invited us to a dinner party that evening, largely to introduce us, not realizing, of course, that we had known each other a very long time. Chet announced to the other guests that he had seen me a few years earlier when we sat together on a hot, stalled plane on our way back from Washington. "She took her jacket off," he said, with a look of glee, as if he wanted the guests to understand that he had seen the buttoned-up academic they saw before them in another guise. He had seen her with arms (and feet) revealed. "I had been at an Oxfam board meeting," he explained, "while she was lecturing spooks at CIA."

After dinner I offered Chet a ride. I offered him a ride because I had a car and he didn't, and he was an old family friend. However, he chose to interpret this offer of a ride in a different way.

He called me the next morning. He had returned to Martha's Vineyard, where he was staying with friends, and he wanted me to come out. Not a chance, I told him. I was taking Evan to the playground, and I had work to do. But he knew precisely how to persuade me. It was the Fourth of July, a long weekend.

"How could you do this to your son? Keep him cooped up like that on a long weekend, when you have an offer to go to the Vineyard?" It now seems remarkable to me that at that stage of my life, I still thought of an hour's drive to the ferry as an extremely inefficient use of time. But something about Chet's familiarity made me feel safe, and I was persuaded. I packed up quickly. I did not change my worn-out clothing. After all,

Chet was an old family friend. In half an hour Evan and I were on the road, on our way to the ferry.

Evan raced around the boat, while I raced behind him. He was excited by everything—the waves. The spray of the sea. The gulls hovering above us, coasting—almost without movement—on the gusts thrown up by the motion of the boat. We tossed the remains of his hot-dog bun into the air, and the gulls caught the bread, mid-flight.

But the rolling sea sickened me. People made too much noise, ordering their hot dogs and Cokes. During this phase of my life, for reasons I did not comprehend, I had cut out most forms of stimulation. Being a working single mother was stimulation enough. I rarely saw films. I did not watch television. I rarely traveled more than twenty minutes from my home, unless I was traveling for work. Traveling would send me into what I now recognize as dissociation. Traveling dislodges you; that's why we do it. I wanted to be lodged. I wanted to be planted into the earth.

Even so, I remember a surprising sensation as the ferry came into shore—a sensation of coming home. It must be because I was going to visit someone from my hometown, someone who knew me, or at least knew of me, through several different phases of my life, and might understand why I found it hard to connect the person that I was then with the person I became.

Some people's lives seem to flow in a narrative; mine had many stops and starts. That's what trauma does. It interrupts the plot. You can't process it because it doesn't fit with what came before or what comes afterward. A friend of mine, a soldier, put it this way. In most of our lives, most of the time, you have a sense of what is to come. There is a steady narrative, a feeling of "lights, camera, action" when big events are imminent. But trauma isn't like that. It just happens, and then life goes on. No one prepares you for it.

Chet met us at the boat in a rented convertible. Evan said to him, "You're not as fat as I thought you would be," as if meeting someone he had been waiting for.

After our last meeting, in which my father spelled out his concerns about our interview and I listed some proposed additions, he sends me a letter. He would like to see and approve of my additions, to make sure the reader will understand the actual situation in my family when my sister and I were raped, and the constraints my father faced.

I send him the material. I tell him that I have added the context now, and that I intended to do so in any case, even before he requested it.

And then, whence I do not know, I hear these words come out of my mouth: "This is a story, Dad, about how trauma affected both you and me as well as our relationship. The reader needs to see how we resolve the situation. I think we should apologize to each other, to show how we have reconciled ourselves to our own actions and each other."

Not only was this rude, it was manipulative and not entirely honest. Why did I say it?

There is a pause, as if my father were contemplating his daughter's lack of social graces and considering how that might have occurred. But my father surprises me. "I get it. It's a story that requires resolution."

And then he adds, "That's fine."

I have to press my father to meet with me. I drive myself out there on a day that I know he is free.

We sit in the garden. "I like to be outside," my father tells me.

I do, too, but I don't tell him that. At this point in this long conversation between my father and me, I can focus on him and nothing else.

"The most important thing I can say about why I didn't come back is that I felt that we were estranged. I felt that I couldn't help you," he says.

I don't like this word *estranged*, which reminds me of *stranger* and *strange*. I must have appeared very strange back then, in my father's eyes. I know that. It hurt then, and it hurts now.

"I remember thinking, There is no point rushing home," my father says, "because you will reject me in any case. You still wouldn't talk to me."

I note the present tense of "will reject me." I recall my father's earlier words about his fear of losing us. "Losing one's child is a terrible thing," he said. He wanted me to talk to him, but I couldn't talk unless he made clear that he wanted to listen. I needed his permission to feel, and that permission wasn't granted. I needed him to feel with me in order to feel safe. I understand that now. But I don't say anything. I sit quietly, waiting for more.

"I had no idea what rape was," he says. "I first learned about it reading about yours. I don't understand the mechanics. The whole subject is repellent to me. You never talked to me about it. I reported to the police that you seemed to have gotten over the rape because you weren't saying anything about it. It didn't come up in conversation.

"It was the same after your mother died," he says. "You didn't talk."

What is happening here? The topic of my mother has been taboo in my father's house for more than forty-five years, but now my father is bringing it up again and again, as if a dam has been broken.

We return to the topic of how my grandfather insisted that my mother undergo what my father considered to be "quack chemical procedures" after she was clearly dying. This is the one topic regarding my mother's death that my father has always been able to talk about. It seems to haunt my father—this mad scientist side of my grandfather.

"She was sent home to die. She was in a hospital bed on that porch, that front room," my father says. I never knew this.

We are sitting on a porch now, my father and I, on the cool side of his house.

"Where Grandma used to sit?" I ask, trying to picture my mother dying in that sunny room. I had always imagined her dying in a darker, more secret place. I have a vague recollection from that dark period of my mother lying on her back, with her knees bent, in the guest bedroom. I decided this was a very grown-up position and tried to emulate it. My grandmother sat in that sunny garden room all day long during the last decade of her life. She was obsessed with the news. She read all the papers and then listened to the news most of the day.

"Yes," he says. "She never said she knew she would die, but she must have known. You would try to climb up on her bed, and she would turn her face away and push you off the bed."

She was twenty-eight. She must have thought that this was best, to turn her face away from us, to push us off the bed. Why is my father telling me this now? It is almost too much, this sudden shift, like a mountain stream in the spring.

Somehow, I manage to find the courage to ask, "Did we cry?"

"I don't recall. You were numb. Or you seemed numb; you were quiet. You were like that for a long time. Your face was closed. Sara was just a baby then. She didn't understand what was happening."

Although I am astonished that my father is speaking so freely about my mother's death, I cannot take this in now. Compared with this image of my dying mother pushing me away, the topic of my rape seems familiar. Something that by now, I have processed. I will feel about this later. I want to return to the topic of my rape and what happened afterward.

"I am very sorry that I was so awful when I was a teenager," I say.

"I was really sorry, too," he says. "You kept attacking me."

Now I feel slapped. But I can't help laughing. My father is a very strong and stubborn man.

"Do you want to apologize for not coming back to us right away?" I ask my father.

"I want to explain something to you," he says, suddenly pedantic. "There is nothing for you to forgive and nothing for me to apologize for. Relationships within a family are based on trust."

I'm not sure what he means by this. I wait for more.

"I kept calling Sidney," he says, referring to our family doctor. "He said that you were quiet and tranquil."

"We were on drugs," I scoff.

"Presumably you were on tranquilizers," he says, patient, at least at the moment, with his intransigent daughter. "I kept calling Sidney. . . . We must have had four or five telephone conversations," he says, defensive now.

It was so expensive back then to make overseas calls. It is hard for me to imagine my fiscally conservative father phoning from Norway to the States. He must have been more worried than he has let on.

"Sidney was my eyes and ears. He was giving me up-to-the-minute news. He was my closest friend, and he was there looking out for you. I don't know what anyone could have done for you, beyond that." He did not speak to Lisa at all during that period, he says.

I wonder if my father is actually remembering what happened, or imagining what he thinks must have happened.

"Even if I had tried to come home right away, I would only

have arrived twelve hours earlier," he says. He has said this before, even despite what he knows that the police said, and despite what he wrote to me twenty years ago, when he explained that it was important for him to finish his work there. But I don't believe he is lying. He does not remember anymore. He cannot bear to remember. I understand this.

"This is not consistent with what you've just told me about those four or five calls to Sidney," I say, annoyed now by his denial. Annoyed, even more, by myself and my inability to let go. But I am not going to let him off the hook. I can forgive him; I forgave him long ago. But I will no longer absorb the impact of his denial. And there is more: I want to relieve my father of a pain that he insists that he doesn't feel.

I do not believe in "forgive and forget." To forgive in the truest sense, we must remember first and then forgive, even in regard to ourselves.

Now he concedes, "It's not that I admire what I did. But I remember what I was thinking at the time. If I got there and you rejected me, I would have felt terrible. I suppose that I was protecting myself," he says.

His honesty is piercing my heart.

"You might say I'm really sorry that you were so hurt and for the role I played in it," I propose to my father, at the risk of annoying him still further with this refrain.

"I'm sorry that circumstances exposed you to all that hurt." He is choosing his words carefully now. "I wish I could have changed that, but it was beyond my power.

"If you want me to say I'm sorry about what happened after your rape, I don't feel sorry," he says, with some finality now. "I did what I could. I was following up on you through Sidney. Until I showed up. When I did show up, I was unable to help you."

But then my father adds a heartbreaking summary of his reasons for not returning to us right away: "I thought having to deal with me as well as the rape would have put an added burden on you."

So here we are, my father and I, talking freely about my mother's death and about my rape, as if they weren't taboo subjects. This feels like the end of an age. The end of an age of denial. My feet can finally settle, safely now on the ground.

Now I read out loud to my father the pages from the beginning of this chapter. I look up periodically to check on his reaction. I see what looks like rapt attention on his face, as if I were reading him a fascinating story. I am puzzled.

"Can you hear me?" I ask, uncertain as to whether my father put his hearing aid in. It is a warm day. We can't see the bird feeder from here, but you can hear the birds singing. Perhaps he is listening to birdsong.

"Yes!" he says, with a look of something like guilty pleasure. So. My father is enjoying the story.

When I get to the part about our mountain-climbing trips, I pause to say, "So you see! Those trips meant as much to me as they did to you." He cannot hide his pleasure, though he would like to.

He says, "Such flights of fancy!" He is the engineer, and I am the daughter. But when he hears that we children found the scent of his sweat comforting, he scoffs, "Oy vey."

Now that we've dispensed with this rape and trauma business, I have a more important question for my father.

"Do you believe in God?" I ask him.

Some families talk easily about life's most important questions. Mine isn't one of them. But a terrible weight is lifted between us now, and I feel I might ask my father anything.

"The things that are inexplicable are what I call God. Certain aspects of creation. What caused the big bang? What came before? Some features of evolution are incredible. Look around you. It's so damn beautiful," he says.

I take in the trees beyond our window, the emerald shade.

"And the more we learn about genetics, the more we realize that all life is related. Every person is seven times removed. That tree is about ten times removed from me genetically!" he says, pointing to a white oak.

It is indeed a beautiful tree.

"And so. When I think about it, I can't believe this happened by accident. In some other time in some other place within the universe, the germ of life was begun. How? I can't believe it's just on earth. I think life can move among the stars in the form of spores that can outlast all kinds of environmental abuse."

I want to bring us back to the earth.

"What about the Nazis?" I ask. "Didn't living through that period in Germany make you question the existence of God?"

"I don't believe that there is an old man with a beard like that painting by Leonardo da Vinci. My mother's God does not exist. God is a life force, the prime mover that created a whole world of possibilities, including Mozart and including the Nazis. I don't know if 'He' was conscious of what he created." My father indicated that "He" should be in quotes.

And then he adds, with an unusually fatherlike tone, "It's even worse what happened in Rwanda. Stay away from war zones. Stay away from soldiers."

His warning about war reverberates in my mind. I have a premonition that I will remember his words.

People often ask me if studying violent people has led me to lose faith in human nature. It hasn't. On the contrary, when you see all the terrible things that people persuade themselves they are doing for the good of humanity, or to right some terrible wrong, it makes you appreciate the possibility of good even more.

I've heard it said that there is no faith without doubt. I will confess to many doubts. My faith has been tested time and again. I am not the sort of person who believes that evil doesn't exist. I know that it does.

It's harder to have faith in God when strange, senseless things seem to happen to you or your family. Why did my grandfather irradiate my mother? Why did the radiation kill her? God does not play dice! These questions—Why do bad things happen to good people? Why do good people do bad things?—have come up repeatedly in my life.

Philosophers traditionally identify three kinds of evil: moral evil—suffering caused by the deliberate imposition of pain on sentient beings; natural evil—suffering caused by natural processes such as disease or natural disaster; and metaphysical evil—suffering caused by imperfections in the cosmos or by chance, such as a murderer going unpunished as a result of random imperfections in the court system. The use of the word *evil* to describe such disparate phenomena is a remnant of pre-Enlightenment thinking, which viewed suffering (natural and metaphysical evil) as punishment for sin (moral evil).[7]

It seems to me now that there is a spectrum between these forms of evil. Cancer is usually thought of as natural evil. But what about my mother's death? My grandfather believed in the health-giving properties of massive doses of radiation. When he irradiated my mother, he thought he was protecting her. But it turned out he was killing her. A kind of zealotry led him to maintain an X-ray machine at home and then use it repeatedly on his own daughter.

What about the evil of rape? What if the perpetrator is mentally ill? What if he has been severely traumatized, as was the case with my rapist? Psychoanalysts believe that when the pain of trauma is so great that the victim cannot sustain feeling, the victim becomes susceptible to preying on others.[8] In this case, the suffering of trauma can lead to the sin of violence, rather than—as pre-Enlightenment philosophers believed—sin leading to suffering.[9] Thus, violence might sometimes be a mixture of natural evil, arising from disease, and moral evil. I am not suggesting here that we not hold perpetrators fully responsible for their crimes. We must act as if they were, in order to prevent further violence.

What about the evil of terrorism? What of the evil of war? Absent intervention, victims of torture or terror or war may raise tortured children who, in turn, are more susceptible to harm their own children psychologically.[10] Male children raised in cultures of violence are more likely to become delinquents or violent criminals.[11] For Jung, evil was inherent, not only in every human being, but also in God. He viewed evil as an archetypal Shadow, an aspect of the unconscious that cannot be controlled, but can be integrated. When it is integrated, it becomes a source of creativity. When it is repressed, it can lead to overt acts of evil. All of these approaches to evil seem to me to be important for comprehending the monstrous acts of a man like Brian Beat, and the difficulty people have believing that he was guilty.

"What do you think happens after death? Are you going to visit me after you die?" I ask my father, conscious of an embarrassing, childlike longing, which I am finally able to reveal.

"After you die, you are in the minds of others, in the perturbations and resonances that you created in society while you were alive. Societal evolution is a thousand times more rapid than the evolution of creatures or life. We have an opportunity to modify social structures according to our values. Talking to you is a way of accomplishing some of that, for example. Talking to Evan. I did that through my work."

And then he concludes, "Having the opportunity to contribute is our most sacred job," he says. I agree completely. I believe he has. I hope we have.

Postscript

The process of writing this book has taught me a great deal about the lingering effects of severe trauma. It is important to point out that no two victims of trauma will have precisely the same symptoms. But some symptoms are common among people suffering from post-traumatic stress disorder. One is difficulty accepting love, or trusting others to take care of you. Another is the sensation of numbness. Another is difficulty recognizing the feeling of fear.

For each individual who has suffered extreme trauma, there will be specific triggers that alter one's receptivity to inputs—causing hypervigilance, on the one extreme, and hypovigilance, a frustrating feeling of calm, on the other. I fear that I may always be subject to these altered states. Hypervigilance makes it possible rapidly to scan one's surroundings for any kind of threat. I am able to react extraordinarily quickly, while feeling relatively calm and in control. In this state, I become efficient and can accomplish many tasks, but I am likely to be unintentionally curt and rude. It can feel like a kind of high, as Skip Shea explained. One can become a "prisoner of detail," as Skip described it, as I was when observing the customers in the gas station after I interviewed Erik. Afterward it is shattering. You feel "physically and emotionally low," to use my father's words, as if you had been dropped halfway into your own death. Sometimes I experience a calm so deep that I cannot focus. It is as if I were underwater. It is almost impossible to drive when I am in this state; I can get lost in places that I know very well.

These altered states can be useful if you know how to harness them; but they can also be quite debilitating. The confusion and pain of transitioning from hyperarousal to an almost complete absence of feeling can make victims turn to drink or drugs or promiscuous sex. In extreme cases, victims are susceptible to suicide or violence against others. A girl who was raped right after my sister and me, probably by the same rapist, killed herself soon afterward.

The soldiers returning from the wars on terrorism are especially prone to suffer symptoms of post-traumatic stress. Because they are fighting in cities, the soldiers are more likely to cause or encounter civilian deaths. And because improved medical technologies have made it possible to keep severely wounded troops alive, they are returning with more extreme injuries and more horrifying memories. In the absence of effective treatment, it is likely that some of them will be vulnerable to alcoholism, drug abuse, and violence—against themselves and others—for many years to come.

Victims of trauma often suffer flashbacks in the form of violent nightmares or puzzling reactions to triggers, such as sounds or scents. It can take a very long time to recognize these triggers. It could be something as mundane as the scent of tuna fish, for example, which might bring back the feeling of terror, if not the memory, of being shot in the leg while eating a sandwich.

Some of the symptoms of post-traumatic stress disorder can be positive in some situations. It is possible to experience what psychologists call post-traumatic growth. The ability to stay calm when endangered can be a tremendous advantage. Survivors are sometimes able to contemplate painful truths that other people prefer to deny, such as hidden malevolence or dangers. This, too, can be an advantage, although it can be annoying to people who prefer to look away. There can also be a heightened awareness of the fragility (and preciousness) of life—one's own and the lives of others. People who have been repeatedly exposed to severe trauma can contribute greatly to society, for example, by joining humanitarian missions in dangerous places or by serving in the Special Forces.

In my case, I know that I will never be "cured." I understand that I am at risk of overprotecting my son or, conversely, exposing him to unnecessary danger or ignoring his needs. I am also at risk of trying to train him to survive dangers he is unlikely to encounter outside my imagination. The goal, it seems to me, should be to learn to manage one's symptoms—to learn techniques for remaining in the present, not just in one's thoughts but also in one's feelings—even when there is no danger or urgency to fix one's gaze. To recognize that danger or urgency—which makes PTSD sufferers feel calm—can become addictive. To learn to distinguish one's reactions to "then" from reactions to "now." To recognize triggers and one's reactions to them, and to use them as clues about how to create a meaningful life.

Like many people with PTSD, I strongly resisted the diagnosis, which I considered to be a fad, and strongly resisted treatment as well. Because of my professional interest in national security affairs, I knew about soldiers returning from war with PTSD. It seemed patently absurd to me that a victim of sexual or relational trauma would suffer the same physiological effects as a soldier returning from war. It took a long time for me to accept the idea that treatment could help me. Once I surrendered to treatment, the results were not entirely positive. I got a lot more accomplished before undergoing treatment. I was a whirling dervish, extraordinarily energetic much of the time. The only problem was that I couldn't sit still. Sustaining that kind of pace precludes intimacy, even with one's children. At some point, intimacy began to feel more important to me than efficiency. But for many people who suffer PTSD, the tradeoff might not be worthwhile. I won't pretend that there aren't grave losses if you choose to be treated for post-traumatic stress disorder, even if, in my case, there were also many gains.

Here is what is most important in my case: I have learned to recognize the sensation of fear. I have learned, I hope, how to love.

Now I want to go out on a limb and propose some hypotheses about possible connections between the two things I now know the most about—terrorism and terror. At this point I am presenting intuitions, rather than conclusions.

I have been researching and writing about terrorism since the mid-1980s. As part of this work, I have interviewed religious-extremist terrorists in America, India, Indonesia, Israel, Lebanon, Pakistan, and Palestine. There were many differences among the terrorists I interviewed. Some were intellectuals. Some appeared to be on a spiritual high; others seemed pumped up on adrenaline or the adventure of living at least partly on the run. Some clearly enjoyed their status and power. Sheikh Fadlallah, the spiritual leader of Hezbollah who was said to have survived an alleged CIA-led attempt on his life in 1985, exuded the air of a man who feels he has the best possible job in the world.

Some were obviously angry. I sensed in many that their commitment to the cause was a thin veneer covering some deeper, more personal need. For some, jihad had become a high-paying job; a few admitted they would like to quit but couldn't afford to. Some were unexpectedly rich, while others lived in slums. Some spoke of grievances that were widely held in their societies, while others had complaints that were not widely shared.

While the terrorists I met described a variety of grievances, almost every one talked about humiliation. An Identity Christian cultist told me he suffered from chronic bronchitis as a child, and his mother discouraged him from exerting himself. He had been forced to attend the girls' physical education classes because he couldn't keep up with the boys. "I don't know if I ever got over the shame and humiliation of not being able to keep up with the other boys—or even with some of the girls," he said. The first time he felt strong was when he was living on an armed compound, surrounded by armed men.

A man involved in the violent wing of the antiabortion movement told me he was "vaginally defeated," but now he is "free," by which he meant celibate and beyond the influence of women. A Kashmiri militant founded his group because he wanted to re-create the golden period of Islam, "to recover what we lost. . . . Muslims have been overpowered by the West. Our ego hurts . . . we are not able to live up to our own standards for ourselves."

The notion that perceived humiliation could be an important factor in explaining terrorism struck some of my colleagues, at least initially, as far-fetched. But my argument is not that humiliation alone is sufficient to create a terrorist. My hypothesis is that humiliation is a risk factor for terrorism.

And then there is the question of rape and torture. Why did interrogators in Iraq and at Guantanamo employ rape and sexual torture? Is it possible that they were exorcising their own shame, even as they believed that sexual torture was a necessary means for extracting information from people they believed to be terrorists?

It is only after commencing the research I describe in this book that I realize the possible importance of the frequency of rape at the radical madrassas I studied in Pakistan. Sexual abuse of madrassa students is widely covered in the Pakistani press, but rarely discussed in the West. I have felt, in my interviews of terrorists, that there was an element of sexual humiliation, but it was rarely more than an intuition, and I have never explored this issue.[12] Also troubling is the rape of boys by warlords, the Afghan National Army, or the police in Afghanistan. Such abuses are commonplace on Thursdays, also known as "man-loving day," because Friday prayers are considered to absolve sinners of all wrongdoing. David Whetham, a specialist on military ethics at King's College in London, reports that security checkpoints set up by the Afghan police and military have been used by some personnel to troll for attractive young men and boys on Thursday nights. The local population has been forced to accept these episodes as par for the course: they cannot imagine defying the all-powerful Afghan commanders. Could such sexual traumas be a form of humiliation that contributes to contemporary Islamist terrorism?

Aside from the question of preexisting personal trauma, consider the impact of a terrorist's lifestyle on his psychology. Exposure to violence, especially for those who become fighters, can cause lasting, haunting changes in the body and the mind. Terrorists are "at war," at least from their perspective, and like soldiers, they, too, may be at risk of post-traumatic stress disorder. Those who were detained may have been subjected to torture and left with even more serious psychological wounds. A number of governments are attempting to "rehabilitate" low-level terrorists. For example, the U.S. military has been attempting to rehabilitate detainees held in U.S.-controlled detention facilities in Iraq. It will be critically important to incorporate some of what the medical community is learning about PTSD in these efforts, not because terrorists deserve sympathy, but because understanding their state of mind is necessary to limiting the risk that they will return to violence.

As I write this book, thousands of soldiers are returning from Iraq and Afghanistan. Most of these soldiers will be affected by the experience of combat, both physically and mentally. Studies suggest that the majority of soldiers recover, psychologically, within several months of their return.[13] Some returning veterans—perhaps many—will experience what psychologists call post-traumatic growth.[14] The army is working together with psychologists to develop a training program to improve combat veterans' emotional resiliency, with the aim of improving combat performance as well as reducing the frequency of post-traumatic stress disorder and suicide.[15]

An estimated 20 to 30 percent of those who return report lingering symptoms of post-traumatic stress disorder or major depression, a significantly higher percentage than was reported in previous wars.[16] There are many possible reasons for the higher rate of reported cases. Deployments have been longer, and breaks between deployments are less frequent than in previous wars.[17] The pace of deployments is unprecedented in the history of the all-volunteer force.[18] Redeployment is a major risk factor for PTSD.[19] At the same time, advances in medical technology and in body armor have reduced casualty rates.[20] Soldiers are surviving combat situations that would have killed them in the past, but returning with traumatic brain injuries and with the memory of mind-breaking horrors. The difficulty of persuading soldiers to redeploy is putting enormous pressure on recruiters, who are increasingly overlooking known mental health problems, increasing the risk that troops will return with more severe psychological illness. All of these factors have increased the incidence of PTSD. The impact of soldiers' untreated despair and hypervigilance will be borne, not only by the soldiers and their families, but also by society at large.

When we train soldiers for battle, we deliberately inculcate in them qualities that, when they return, we will refer to as symptoms of PTSD.[21] Soldiers must be able to dissociate, to cut off emotion. A soldier who collapsed in tears because one of his comrades was killed or because he saw the remains of a shattered baby on the sidewalk would put his own life and the lives of others at risk. Soldiers must be able rapidly to scan their environment and respond immediately to threats. These qualities are components of what we often call strength and courage. They are the qualities that keep soldiers alive and allow them to protect their comrades.

Numbness and hypervigilance can keep you alive when you're literally under the gun. They can even make you more efficient when you are under the gun metaphorically. But these very same qualities, necessary to keep us alive when we are threatened with death, get in the way of normal life and of human relationships. The same hypervigilance that keeps a soldier alive can make him throw himself on the floor when he hears a car backfiring. For the most traumatized soldiers, sounds or situations that trigger emotions they don't recognize or understand can lead them to hurt themselves or others.

In the moment that a person's life is threatened, the separation of thought and feeling may be necessary to sustain life. But this separation can become a habit, and when it does, we are only half alive. This book is a memoir—not of specific life events, but of the processes of dissociation, and of reenlivening emotions that are shameful to admit or even to feel. It is an account of the altered states that trauma induces, which make it possible to survive a life-threatening event but impair the capacity to feel fear, and worse still, impair the ability to love.

My goal in writing this book is to help not only the millions of women and men who have been raped or tortured but the soldiers who risk their lives on our behalf, returning with psychic wounds so excruciating that both they and we cannot bear to admit that these wounds exist. Denial is almost irresistibly seductive, not only for victims who seek to forget the traumatic event but also for those who observe the pain of others and find it easier to ignore or "forget." In the long run, denial corrodes integrity—both of individuals and of society. We impose a terrible cost on the psychically wounded by colluding in their denial.

notes

1. Catharine A. MacKinnon, "Women's September 11th: Rethinking the International Law of Conflict," Harvard International Law Journal 47, no. 1 (Winter 2006): 25.
2. The "walking corpses" is Bruno Bettelheim's term in *The Informed Heart* (New York: Free Press, 1960), p. 151. Psychiatrist and author Henry Krystal "affirms that psychogenic death can occur if the victim of catastrophic trauma completely surrenders to the situation in which no action is perceived as possible. If this surrender occurs, he/she falls into a state of immobility (catatonia), and abandons all life-preserving activity. He calls this a 'potential psychological "self-destruct" mechanism' and affirms that, once the process of total surrender starts it is no longer voluntarily terminable but may only be stopped by the intervention of an outside caretaker, and that, if this does not happen, the victim will die." Krystal cited in Carole Beebe Tarantelli, "Life within Death: Towards a Metapsychology of Catastrophic Psychic Trauma," *International Journal of Psychoanalysis* 84 (January 2003): 915–28.

3. Primo Levi, *The Drowned and the Saved* (New York: Simon & Schuster, 1988), 79; cited in Tarantelli, "Life within Death."
4. Feminists argue about whether rape is a form of sex. See, for example, Susan Brownmiller, *Against our will: Men, Women, and Rape* (New York : Simon and Schuster, 1975); Catherine A. MacKinnon, "Sexuality, Pornography, and Method: 'Pleasure Under Patriarchy'" Ethics, Vol. 99, No. 2 (Jan., 1989), pp. 314–346; and Camille Paglia, *Vamps and Tramps: New Essays* (New York: Viking, 1994). It didn't feel like sex to me, maybe because the gun and the imminence of death were so central to my experience.
5. http://www.thedoctorwillseeyounow.com/articles/behavior/ptsd_4/
6. Tim O'Brien, *The Things They Carried* (New York: Houghton Mifflin, 1990).
7. Gottfried Leibniz's answer to the question of why God would allow a natural order that involved so much innocent suffering was that man brought such natural evils upon himself: natural evil was collective punishment for moral evil, including but not limited to the Fall. A massive earthquake that destroyed the city of Lisbon in 1755 evoked a reaction among Enlightenment philosophers and theologians similar to that of their twentieth-century counterparts to Auschwitz, Susan Neiman explains in *Evil in Modern Thought: An Alternative History of Philosophy* (Princeton, N.J.: Princeton University Press, 2002), 1–57. Jean-Jacques Rousseau would reject Leibniz's view, ushering in a more modern conception of evil. Innocent suffering was not punishment for sin, but a symptom of ignorance. In regard to the earthquake at Lisbon, for example, it made no sense for humans to live in large cities where they were vulnerable to earthquakes. Interestingly, psychiatrists are seeing new links between suffering and sin today, as we shall see.
8. Sue Grand, *The Reproduction of Evil: A Clinical and Cultural Perspective* (Hillsdale, N.J.: Analytic Press, 2000). Victims of repeated abuse, or children who live in violent neighborhoods or war zones,

may experience PTSD; see Tener Goodwin Veenema and Kathryn Schroeder-Bruce, "The Aftermath of Violence: Children, Disaster, and Posttraumatic Stress Disorder," *Journal of Pediatric Health Care* 16, no. 5 (September–October 2002): 235–44.

9. Shakespeare's Richard III, for example, attributes his determination to become "a villain" (morally evil) to his having been "cheated of feature by dissembling nature" (natural evil). This might be a good example of how victimization—including by fate—can be used to justify moral wrongs. Derek Summerfield observes that "the profile of post-traumatic stress disorder has risen spectacularly, and it has become the means by which people seek victim status and its associated moral high ground in pursuit of recognition and compensation." Summerfield, "The Invention of Post-traumatic Stress Disorder and the Social Usefulness of a Psychiatric Category," *British Medical Journal* 322 (January 2001): 95–98.

   Judith Lewis Herman, a leading authority on trauma, reminds us that it is important not to subscribe to the belief that traumatized victims inevitably become evil, since that would involve "blaming the victim," which is morally wrong. Herman points out that there is a popular literature of the "cycle of abuse" theory. Herman states that this theory is not empirically valid, explaining that "its most glaring weakness is its inability to explain the virtual male monopoly on this type of behavior. Since girls are sexually victimized at least twice to three times more commonly than boys, this theory would predict a female rather than a male majority of sex offenders." Herman, "Considering Sex Offenders: A Model of Addiction," *Journal of Women in Culture and Society* 13, no. 4 (Summer 1988): 703–4.

10. Some studies suggest that PTSD may be at least partly heritable. Adult children of Holocaust survivors are at greater risk for PTSD—perhaps because they are more likely to expose themselves to traumatic events through life choices, or because susceptibility to PTSD is heritable. See R. Yehuda et al., "The Cortisol and Glucocorticoid Receptor Response to Low Dose Dexamethasone Administration in

Aging Combat Veterans and Holocaust Survivors With and Without PTSD," *Biological Psychiatry* 52, no. 5 (September 2002): 393–403; M. B. Stein, K. L. Jang, and S. Taylor, "Genetic and Environmental Influences on Trauma Exposure and Posttraumatic Stress Disorder Symptoms: A Twin Study," *American Journal of Psychiatry* 159, no. 10 (October 2002): 1675–81.

11. Judith Herman reports that the majority of trauma victims do not become perpetrators, but that trauma appears to amplify common gender stereotypes among victims of childhood abuse. Men are more likely to take out their aggression on others, while women are more likely to injure themselves or to be victimized again. Judith Lewis Herman, *Trauma and Recovery* (New York: Basic Books, 1992), 113. For adolescent males, exposure to violence and victimization is "strongly associated with externalizing problem behaviors such as delinquency, while adolescent females exposed to violence and victimization are more likely to exhibit internalizing symptoms," according to Z. T. McGee et al., "Urban Stress and Mental Health among African-American Youth: Assessing the Link between Exposure to Violence, Problem Behavior, and Coping Strategies," *Journal of Cultural Diversity* 8, no. 3 (Fall 2001): 94–104. A 1999 study showed that male Vietnam veterans seeking inpatient treatment for PTSD were more likely to exhibit violent behavior than a mixed diagnostic group of inpatients without PTSD: M. McFall et al., "Analysis of Violent Behavior in Vietnam Combat Veteran Psychiatric Inpatients with Posttraumatic Stress Disorder," *Journal of Traumatic Stress* 12, no. 3 (July 1999): 501–17. Combat exposure was found to have an independent positive association with interpersonal violence, when controlling for PTSD among combat veterans. F. C. Beckham et al., "Interpersonal Violence and Its Correlates in Vietnam Veterans with Chronic Posttraumatic Stress Disorder," *Journal of Clinical Psychology* 53, no. 8 (December 1997): 859–69. Dr. Jerrold Post argues that PTSD, secondary to living in a Palestinian refugee camp, could play a role in the creation of a terrorist. Post, "Terrorist on Trial:

The Context of Political Crime," *Journal of the American Academy of Psychiatry Law* 28, no. 4 (2000): 489.

12. I will not be able to answer this question here. First, I no longer feel safe interviewing terrorists in the field. Second, even if I did, it would be difficult to pose such a question to terrorists (it would be too humiliating to them). To address the question empirically would require surveying a random sample of terrorists as well as controls. We would need to know the relative frequency of humiliation (sexual or otherwise) in the general population as well as in the terrorist population. It would also be interesting to understand the frequency of post-traumatic stress disorder among terrorists and controls, and to ask terrorists about possible connections between their experience of trauma and their choice of career. This sort of study would be very difficult to carry out. But perhaps someday it will be possible.
13. William Nash, "PTSD 101," course transcript for Combat Stress Injuries, Department of Veterans Affairs, National Center for PTSD. This is an excellent, humane description of the psychological impact of war, available at www.ncptsd.va.gov.
14. Ibid.
15. Benedict Carey, "Mental Stress Training Is Planned for U.S. Soldiers," *New York Times*, August 17, 2009.
16. Lisa H. Jaycox and Terri Tanielian, eds., *Invisible Wounds of War: Psychological and Cognitive Injuries, Their Consequences, and Services to Assist Recovery* (Santa Monica, Calif.: Rand, 2008), available at http://www.rand.org/pubs/monographs/MG720/. The 30 percent figure is from Ann Scott Tyson, "Military Diagnosing More Post-Traumatic Stress," *Washington Post*, May 28, 2008.
17. James Hosek, Jennifer Kavanagh, and Laura Miller, *How Deployments Affect Service Members* (Santa Monica, Calif.: Rand, 2006); Jaycox and Tanielian, *Invisible Wounds*.
18. Amy Belasco, "The Cost of Iraq, Afghanistan, and Other Global War on Terror Operations Since 9-11," Congressional Research Service

report for Congress, June 28, 2007; Jaycox and Taneilian, *Invisible Wounds*.

19. Tyson, "Military Diagnosing."
20. Jaycox and Tanielian, *Invisible Wounds*.
21. Jonathan Shay has written about this issue extensively; see, for example, *Achilles in Vietnam* (New York: Scribner, 1994).

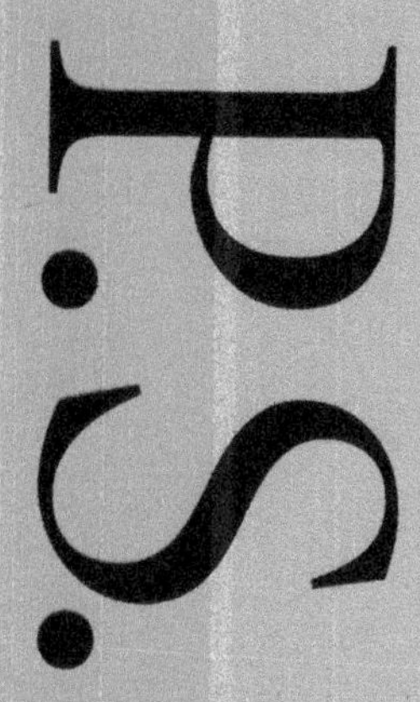

Insights,
Interviews
& More . . .

About the book

Read on

I Am Lucy
by Amy Vorenberg

TERRORISM EXPERT Jessica Stern's *Denial* is a memoir of the 1973 rape of her sister and herself. By itself, the story of the dual rapes is horrendous, but it is more appalling because they were probably preventable.

Jessica and her sister were victims of a man who raped or attempted to rape at least forty-four girls in the Boston area from 1970 to 1973.

I was number twelve. In Jessica's book, I am Lucy.

Shockingly, eighteen of the forty-four rapes occurred within an eight-block radius of Harvard's Radcliffe campus in Cambridge, many at Radcliffe itself. Yet there was barely a whisper of these crimes in the media. While some short articles appeared in a campus newspaper, there was no systematic attempt to alert the community at the time, even though the rapist followed an obvious pattern. All the rapes occurred between 8:30 P.M. and 10:30 P.M. All the victims were under the age of nineteen and gave similar accounts of their attacker's age, build, eyes, and hair. The rapist carried a small, white-handled gun and wore a ski mask. And he often chose a women's dormitory or a house where several girls lived.

I was thirteen in 1971 and lived just a few blocks from the Radcliffe quadrangle. Our house was always full of visitors, so it was no surprise that on an unseasonably warm March night there were eleven people at my home. What we didn't know, as we sat and talked in our kitchen, was that a lone man had been on a sexual assault spree in our neighborhood for most of the preceding year.

At 8:30 P.M. I went upstairs to take a shower. As I came out, there was a man in the bathroom. He carried a gun and wore a mask. Apparently he had come up a fire escape and through an unlocked window. Frozen and scared, I did what he demanded.

Four years ago, I decided to learn anything I could about the rapist. During my search for information, I crossed paths by chance with Jessica Stern. Jessica had retrieved her police file, and a diligent detective took it upon himself to investigate her decades-old case. He assembled evidence from many towns showing that dozens of unsolved rape cases from Cambridge to Natick fit the same pattern. No DNA evidence was collected back then, but the description of the perpetrator, his gun, and his distinctive MO left little doubt that the crimes were the work of the same man.

Enduring rape at thirteen was rough. But to realize now that I was one of forty-four is just hard to process—a fact made harder by the knowledge that my sisters and I were essentially sitting ▸

> “Jessica and her sister were victims of a man who raped or attempted to rape at least forty-four girls. . . . I was number twelve. In Jessica's book, I am Lucy.”

ducks. Having heard nothing from the police or the university, my parents had taken no extraordinary steps to protect their daughters.

Although forty years have passed, respected institutions still suppress information about sexual assault, and rape remains the most underreported of violent crimes. In February, the Center for Public Integrity reported findings of a twelve-month investigation of campus rape. It found that colleges and universities continue "ducking bad publicity" by misrepresenting statistics on campus rapes.

In my case, Harvard/Radcliffe was guilty of failing to protect its students, as well as girls like me in the surrounding community. But the institution's withholding of information demonstrated something even more insidious that persists today.

Rape is a violent act of forced submission. Institutions that condone silence are complicit in perpetuating the shame associated with rape, confirming the message to victims that they should hide their shame and suffer in silence. I have been silent long enough.

"Institutions that condone silence are complicit in perpetuating the shame associated with rape.... I have been silent long enough."

Amy Vorenberg is a professor of law at Franklin Pierce Law School at the University of New Hampshire. This essay, courtesy of the author, has been adapted from an article that originally appeared in the Boston Globe *on June 20, 2010.*

Readers Respond to *Denial*

The most moving aspect of the publication of Denial *has been the response from readers. I've heard from hundreds of readers who had been raped or experienced other kinds of trauma who wanted to share their experiences with me. Some of them agreed to have portions of their letters printed here, in their own names or with pseudonyms. Several of the women who wrote to me had also been raped by Brian Beat. Like me, they had no idea, until they read* Denial, *that their perpetrator was a serial rapist. I hope that these letters provide comfort to readers who have experienced trauma.*

January 19, 2011

Dear Ms Stern,

I've just read *Denial*, and wish to thank you so much. I'm a fifty-eight-yr-old woman who has waited all my life for someone to affirm & confirm my experiences as a survivor of child sexual abuse.

It was apparent on almost every page that it took great courage to write a book with such raw honesty. Several times, upon beginning to read, I almost put it away—but I pushed through, somehow knowing the ultimate benefit would be worth my discomfort.

It's impossible to express the relief I ▸

feel now, having thrown off the self-doubt which is the spoiled fruit of denial—I didn't "imagine or fabricate" any of what I experienced.

While I am so very sorry for what happened to you—the rape, & the subsequent denial—I'm very grateful to you for writing the book &, in effect, telling me, "yes, I know—it is true, & it was not your fault." The concept of "blaming the victim" is one I'll never be able to wrap my head around—however, you did an excellent job of explaining that much of it comes from people not wanting to feel pain (the victim's & their own). I can't imagine the strength it takes for you to interview victims—just reading their stories is heartbreaking, almost physically painful (the former altar boy, passed from vile priest to priest).

Sincerely,

N.N.

August 16, 2010

Dr. Stern,

Just finished your book and I am still astounded by your findings. Never in my wildest dreams did I expect that I myself might have PTSD.

I was born in 1957 in Transylvania/Romania into the German minority group and grew up as the only daughter of a government official who owned (and tortured) the three generations of women in his household (mother-in-law/wife/daughter).

I defected in 1983 at the age of twenty-six and experienced from then on most of the symptoms you are describing in your book. Outwardly doing well: Assistant Prof at the University of Maryland/European Division Heidelberg; brought mother and grandmother to freedom (illegally; grandmother wrapped in a tapestry) . . . despite PTSD?! Even tried to write a book about my experiences.

I've been in treatment for depression for a long time . . . Prozac, then Paxil silencing most of my symptoms—but it helps to understand why I might feel/act the way I do, i.e. charity work for my village of birth became almost an obsession.

Thank you for a very well documented, well written and insightful memoir. And I am flabbergasted about not only your honesty, but your willingness to invite us into your life.

You are one of those special women—so intelligent, educated and beautiful—who lifts the rest of us up!

Respectfully,
Lilly Pierce

August 13, 2010

Dear Jessica Stern:

Thank you for having the courage to publish *Denial.* It is one of the clearest accounts of responses to sexual terrorism that I have read in the past twenty years. Of particular interest to me were your clear examples of the focused, intense hyperawareness that can overtake you when you are in an intense work situation.

Your book was like talking to someone who understands post trauma from a depth beyond academic research. An understanding that balances academic study with the deep realizations that can only come from personal experience. I've read just about every book I could get my hands on about surviving rape, war, sexual abuse, emotional abuse, growing up in an alcoholic home, etc. I used bibliotheraphy as a way of understanding and dealing with the events in my early life. Your book is at the top of my list because of the way it explores the consequences of shame and how it ripples through a family.

I know someone who was diagnosed with PTSD after serving eighteen months in Iraq. It is hard to see him struggle in ways that are familiar to me. I was selected as a candidate for a PTSD drug trial for the NIMH a few years back, but I turned it down. ▸

I too had a hesitancy about accepting that survivors of sexual trauma and veteran's experiences are comparable.

The hyperawareness and the intuitive sharpness you accurately describe in your book have helped me in my job tremendously. Though it is often accompanied by the sleepy backlash—parasympathetic backlash, as described by the psychologist Dave Grossman—that comes later.

I conduct simulation-based training to prepare people for events ranging from someone bringing a weapon to work, to someone just giving effective performance feedback. We freeze scenes and analyze the situation, the emotions, and the professional and personal identity factors of each person in the scene. It is intense, focused and sometimes exhausting work.

The skills that help me most are not the ones I learned studying Russian or the ones I learned in law school. They are the ones I learned as a survivor of sexual abuse. The three years spent in SIA (Survivors of Incest Anonymous) meetings also helped teach me emotional and social management skills. I later got certifications in the Emotional and Social Competence Inventory (a 360 instrument for leaders) and in the Myers-Briggs Type Indicator (MBTI); but my early life experiences gave me a survivor's toolkit long before academic study helped me understand what it was. I appreciated your recognition of the benefits and challenges caused by being a professional and a survivor of sexual terror.

Thank you again for what you have done.

M.E. Hart

Aug 9, 2010

Dear Jessica,

I am almost finished reading your book *Denial.* I want to thank you for writing something that finally speaks to my professional and personal experience(s).

I was sexually abused from the ages of twelve through sixteen and I pushed the experiences and memories as far away as they would go. After many years of running, falling down, picking myself up, therapy and finding myself, I now understand who I am and why I do the work that I do.

I have been working with dying children and their families here at SickKids in Toronto for the past twelve years. I also facilitate support groups for women with eating disorders at a centre called "Sheena's Place." And, I work with private clients, mostly women, who have experienced trauma in their early childhoods.

I now know that my engagement in and passion for this work are not coincidences.

I am now forty-one years old, married to a wonderful man who loves and understands me. We have a beautiful daughter who is almost four years old. I strive to find joy in the "here and now"—and some days have it easier at achieving this than others.

Your courage and strength are inspiring.

Thank you, thank you, for sharing your story,

Deborah S. Berlin-Romalis

Aug 3, 2010

Thank you for writing your book. I am on every page (continual sexual assault by teenage brothers) and each page describes me. I wondered why I was such an insightful therapist, knowing and feeling things, giving credit to God for giving me insight. I wondered why I would go numb in some areas of my life, knowing things, or not knowing things, getting lost, etc. And you answered the question others have asked me "Why didn't you stop it?" I had no answer. You answered the question "Why can't I scream or yell if I'm in trouble?" Now I know. I have so many other facets of my life that go unanswered, and you have helped me so much on this path of knowing that I was abused, even though my mom said it was my fault that I was a girl, and even though others took advantage of me over the years. ▸

I skip moments and go into hiding mentally, at times, but each day brings new answers and new hope. I think I am even beginning to understand my anger at things that aren't fair, my anger at the polarization in politics … all of those things are beginning to make sense. Because violence scares me and nobody protected me as a child/teenager.

Terrorism . . . it was in my home. I was afraid to go to my home, after school, because I knew what was going to happen to me. I was so frightened when my brother returned home after a date because I knew what was going to happen. Pages and pages of fearful things and I wonder why I had no words to describe what you described in your book.

You gave me a key to go further.

Fondly,

Barbara Meuleman-Girga

Jul 29, 2010

Dear Dr. Stern,

Last night, I completed reading *Denial*.

In your book, you had mentioned that one of your goals in writing it was to help others with similar experiences. I hope this finds you, to inform you that you have indeed helped me, tremendously.

I am a criminal lawyer. In said capacity, I represent rapists, murderers, drug dealers, gang members, and so forth. I was sexually assaulted when I was young. I have also survived war. I was born in the Middle East, and eventually immigrated to the U.S. My childhood memories are filled with abuse and sounds of war. Consistent with what you describe in your book, I, too, have a complete non-reaction to fear or what most "rational" or "normal" individuals would characterize as threatening situations. I had PTSD for a very long time. Perhaps I still do.

Your book shed light on many things for me—things I had suspected all along, but never truly gained insight into, despite extensive therapy and my strong desire to connect the dots.

I am very grateful for your courage and altruism in having written this book.

Thank you!

Melanie D.

Jul 23, 2010

Dear Professor Stern,

As a Boston-area resident, I've been following the story of the re-opening of the investigation of the string of rapes that your own was a part of, and have silently empathized with your experience. I write now because your story has become more personal to me, in a loosely connected way. In 1973, my father too went to Trondheim in Norway to discuss radar technology, so I assume our fathers must have worked together. I think the company or the project was called Digiplot. I was just five years old, but I still have a postcard with a picture of a troll that he sent me then. During that long trip, my mother was left home alone with me and my younger brother, a baby at the time. I know that separation put a terrible strain on their relationship, and my mother was close to breaking down by the time he got back.

Reading your story, I am aware of just how precarious it could be then for women to be alone—of course it is now as well, but now perhaps we are more aware of the unthinkable, and take steps to prevent it. If I understand all the news reports correctly, back then, the police and media did nothing to warn women that this predator was out there. I ache thinking about my mother's vulnerability. Anybody watching the house could have easily learned that she was alone with two small children. And I ache knowing that even though she was safe, others, like you, weren't. And finally, I ache because that is how recently the stories and experiences of women were something inconvenient and of dubious veracity, hysterical in the root sense of the word.

I wish your father had turned to mine and the rest of the team and said, "Here, you handle this; I've got to get home." ▸

Thank you for bringing the issues of fair warning and what is really a form of terrorism against women and children to the front and editorial pages. Your pain may mean progress for us all, and so I thank you for being able to bear the public scrutiny, and letters like this one from people who perhaps thoughtlessly keep dragging it all up again and again.

Wishing you all the best,

Louisa M.

Jul 21, 2010

Dear Jessica,

Thank you for having the beautiful strength and insight to write *Denial*. I am one of the forty-four victims you write about. I am not "out" about this, though it has informed much of my creative work over the years and is probably quite obvious to some (it takes one to know one). Your book is deeply meaningful to me on many, many levels.

Bravo to you,

Elizabeth N.

Jul 19, 2010

Hi, Jessica,

I'm an RN in California and I have PTSD. I can't thank you enough for writing *Denial*. I have so many of the same symptoms as you do. Many years ago I worked in intensive care/coronary care units, the "battlefields" of health care, and I felt right at home. After many, many years of therapy I still don't feel that justice has been served regarding people who caused me harm. I'm not compensated like war vets. I go to work with my symptoms and oftentimes feel sicker than my patients. However, I'm a professional and take great pride in my work as a nurse. Thank you again for your great work. We have very different backgrounds and yet we experience life in much the same way.

Best Wishes,

Sherri Surdyka

Jul 10, 2010

Ms Stern,

Thank you for *Denial*.

I was abducted, raped and beaten by a stranger in 1976. I was twelve. I have made my way in the world largely disconnected, succeeding in a career where performing well in chaos is rewarded.

The fellow who hurt me used a weapon I thought was a gun, then a knife. When he let me go and I learned it was a carving fork, I was devastated. Who the hell gets abducted at forkpoint? When I read the gun used to intimidate you into compliance was a cap gun, I could sense the shame. We wear it like a scent we cannot wash or weaken.

My rapist (I loathe taking ownership of him and it like that, but it is the truth of it, I reckon) abducted dozens of young girls in New York, was convicted of hurting three, served about eighteen years of a twenty-five-year sentence, was granted conditional release and re-offended. He is out now and I am participating in a restorative justice process in hopes that I might meet him so that I can ask him questions that only he can answer.

I wonder if all children who are hurt by strangers carry with them the desire to know their tormentors. Those who hurt us so deeply that way leave their mark boldly, indelibly.

Like you, I worked so hard to be "not a victim" as I understood that concept through my adolescence. I grew distant even from myself and oh, so cool and free of emotion. You are right about the intimacy . . . violence is such a strange intimacy. It was he who saw me at my most vulnerable. It was he who ▸

saw me when I was preparing myself never to go home again. It was he who saw what I was like when I believed I was going to die. Ouch. That makes me closer to him than I have ever been to another human being. I hate that I am closer to him than any other . . . and closer than I seem capable of being.

I just completed the final chapter today and I understand fully that the deepest wound might in fact be lacking the capacity to love. While I recognize it is compromised, I have high hopes it has not been stolen. I have made my way in the world feeling so very small and alone. Your book is heartachingly brave and I want to thank you from a very deep place for showing me that I am not all alone and completely nuts.

When we were kids, people did not talk about the sexual abuse and assault of children. Hell, they rarely talked about sex. I was admonished against strangers. But the strangers I was ready to repel were the kind who would offer me candy, drugs, or money to go with them. Nobody told me that a grown man on foot asking for directions was more dangerous still because it seemed to me that it was he who was vulnerable. I am sure that kind of thing happened way too often. Why did nobody tell us?

Thank you again for taking what seems to me a much greater risk than meeting with terrorists. Thank you for meeting terror and having the courage to express what it is like. And thank you for suffering this interruption. But I could not let the day go by without trying to find the words to demonstrate my gratitude for your effort, your honesty and your great risk.

In Gratitude,

Lin Robins